MATLAB Supplement to Fuzzy and Neural Approaches in Engineering

Adaptive and Learning Systems for Signal Processing, Communications, and Control

Editor: Simon Haykin

Werbos / THE ROOTS OF BACKPROPAGATION: From Ordered Derivatives to Neural Networks and Political Forecasting

Krstić, Kanellakopoulos, and Kokotović / NONLINEAR AND ADAPTIVE CONTROL DESIGN

Nikias and Shao / SIGNAL PROCESSING WITH ALPHA-STABLE DISTRIBUTIONS AND APPLICATIONS

Diamantaras and Kung / PRINCIPAL COMPONENT NEURAL NETWORKS: Theory and Applications

Tao and Kokotović / ADAPTIVE CONTROL OF SYSTEMS WITH ACTUATOR AND SENSOR NONLINEARITIES

Tsoukalas and Uhrig / FUZZY AND NEURAL APPROACHES IN ENGINEERING

Hrycej / NEUROCONTROL: Toward an Industrial Control Methodology

Beckerman / ADAPTIVE COOPERATIVE SYSTEMS

MATLAB Supplement to Fuzzy and Neural Approaches in Engineering

J. Wesley Hines

A Wiley-Interscience Publication
JOHN WILEY & SONS, INC.
New York • Chichester • Weinheim • Brisbane • Singapore • Toronto

Copyright © 1997 by John Wiley & Sons, Inc.

All rights reserved. Published simultaneously in Canada.

Reproduction or translation of any part of this work beyond that permitted by Section 107 or 108 of the 1976 United States Copyright Act without permission of the copyright owner is unlawful. Requests for permission or further information should be addressed to the Permissions Department, John Wiley & Sons, Inc., 605 Third Avenue, New York, NY 10158-0012.

ISBN 0-471-19247-3 (paper)

Printed in the United States of America

10 9 8 7 6 5 4 3 2

To SaDonya

CONTENTS

PREFACE xi
ACKNOWLEDGMENTS xi
ABOUT THE AUTHOR xii
SOFTWARE DESCRIPTION xii

INTRODUCTION TO THE MATLAB SUPPLEMENT 1

INTRODUCTION TO MATLAB 1

MATLAB Toolboxes 3
SIMULINK 3
User Contributed Toolboxes 4
MATLAB Publications 4

MATLAB APPLICATIONS 5

CHAPTER 1 INTRODUCTION TO HYBRID ARTIFICIAL INTELLIGENCE SYSTEMS 5
CHAPTER 2 FOUNDATIONS OF FUZZY APPROACHES 6
2.1 Union, Intersection and Complement of a Fuzzy Set 6
2.2 Concentration and Dilation 9
2.3 Contrast Intensification 10
2.4 Extension Principle 12
2.5 Alpha Cuts 14
CHAPTER 3 FUZZY RELATIONSHIPS 16
3.1 A Similarity Relation 16
3.2 Union and Intersection of Fuzzy Relations 16
3.3 Max-Min Composition 18
CHAPTER 4 FUZZY NUMBERS 19
4.1 Addition and Subtraction of Discrete Fuzzy Numbers 19
4.2 Multiplication of Discrete Fuzzy Numbers 21
4.3 Division of Discrete Fuzzy Numbers 23
CHAPTER 5 LINGUISTIC DESCRIPTIONS AND THEIR ANALYTICAL FORM 24
5.1 Generalized Modus Ponens 24
5.2 Membership Functions 24
    5.2.1 Triangular Membership Function 24
    5.2.2 Trapezoidal Membership Function 26
    5.2.3 S-shaped Membership Function 27
    5.2.4 Π-shaped Membership Function 28
    5.2.5 Defuzzification of a Fuzzy Set 29
    5.2.6 Compound Values 31
5.3 Implication Relations 33
5.4 Fuzzy Algorithms 37
CHAPTER 6 FUZZY CONTROL 44
6.1 Tank Level Fuzzy Control 44
CHAPTER 7 FUNDAMENTALS OF NEURAL NETWORKS 52
7.1 Artificial Neuron 52

7.2 Single Layer Neural Network 57
7.3 Rosenblatt's Perceptron 58
7.4 Separation of Linearly Separable Variables 65
7.5 Multilayer Neural Network 68

CHAPTER 8 BACKPROPAGATION AND RELATED TRAINING PARADIGMS 71

8.1 Derivative of the Activation Functions 71
8.2 Backpropagation for a Multilayer Neural Network 72
    8.2.1 Weight Updates 74
    8.2.2 Hidden Layer Weight Updates 76
    8.2.3 Batch Training 77
    8.2.4 Adaptive Learning Rate 79
    8.2.5 The Backpropagation Training Cycle 80
8.3 Scaling Input Vectors 81
8.4 Initializing Weights 84
8.5 Creating a MATLAB Function for Backpropagation 85
8.6 Backpropagation Example 88

CHAPTER 9 COMPETITIVE, ASSOCIATIVE AND OTHER SPECIAL NEURAL NETWORKS ... 91

9.1 Hebbian Learning 91
9.2 Instar Learning 93
9.3 Outstar Learning 95
9.4 Crossbar Structure 96
9.5 Competitive Networks 98
    9.5.1 Competitive Network Implementation 99
    9.5.2 Self Organizing Feature Maps 103
9.6 Probabilistic Neural Networks 106
9.7 Radial Basis Function Networks 109
    9.7.1 Radial Basis Function Example 113
    9.7.2 Small Neuron Width Example 115
    9.7.3 Large Neuron Width Example 116
9.8 Generalized Regression Neural Network 117

CHAPTER 10 DYNAMIC NEURAL NETWORKS AND CONTROL SYSTEMS 122

10.1 Introduction 122
10.2 Linear System Theory 123
10.3 Adaptive Signal Processing 127
10.4 Adaptive Processors and Neural Networks 129
10.5 Neural Networks Control 131
    10.5.1 Supervised Control 132
    10.5.2 Direct Inverse Control 132
    10.5.3 Model Referenced Adaptive Control 133
    10.5.4 Back Propagation Through Time 133
    10.5.5 Adaptive Critic 134
10.6 System Identification 135
    10.6.1 ARX System Identification Model 135
    10.6.2 Basic Steps of System Identification 136
    10.6.3 Neural Network Model Structure 136
    10.6.4 Tank System Identification Example 138
10.7. Implementation of Neural Control Systems 141

CHAPTER 11 PRACTICAL ASPECTS OF NEURAL NETWORKS 144

11.1 Neural Network Implementation Issues 144
11.2 Overview of Neural Network Training Methodology 144
11.3 Training and Test Data Selection 146
11.4 Overfitting 149
    11.4.1 Neural Network Size 150
    11.4.2 Neural Network Noise 153
    11.4.3 Stopping Criteria and Cross Validation Training 155

CHAPTER 12 NEURAL METHODS IN FUZZY SYSTEMS 158

12.1 Introduction 158
12.2 From Crisp to Fuzzy Neurons 158
12.3 Generalized Fuzzy Neuron and Networks 159
12.4 Aggregation and Transfer Functions in Fuzzy Neurons 160
12.5 AND and OR Fuzzy Neurons 161
12.6 Multilayer Fuzzy Neural Networks 162
12.7 Learning and Adaptation in Fuzzy Neural Networks 164

CHAPTER 13 NEURAL METHODS IN FUZZY SYSTEMS 171

13.1 Introduction 171
13.2 Fuzzy-Neural Hybrids 171
13.3 Neural Networks for Determining Membership Functions 171
13.4 Neural Network Driven Fuzzy Reasoning 174
13.5 Learning and Adaptation in Fuzzy Systems via Neural Networks 177
    13.5.1 Zero Order Sugeno Fan Speed Control 179
    13.5.2 Consequent Membership Function Training 183
    13.5.3 Antecedent Membership Function Training 184
    13.5.4 Membership Function Derivative Functions 186
    13.5.5 Membership Function Training Example 189
13.6 Adaptive Network-Based Fuzzy Inference Systems 194
    13.6.1 ANFIS Hybrid Training Rule 194
    13.6.2 Least Squares Regression Techniques 195
    13.6.3 ANFIS Hybrid Training Example 199

CHAPTER 14 GENERAL HYBRID NEUROFUZZY APPLICATIONS 205

CHAPTER 15 DYNAMIC HYBRID NEUROFUZZY SYSTEMS 205

CHAPTER 16 ROLE OF EXPERT SYSTEMS IN NEUROFUZZY SYSTEMS 205

CHAPTER 17 GENETIC ALGORITHMS 206

REFERENCES 207

Preface

Over the past decade, the application of artificial neural networks and fuzzy systems to solving engineering problems has grown enormously. And recently, the synergism realized by combining the two techniques has become increasingly apparent. Although many texts are available for presenting artificial neural networks and fuzzy systems to potential users, few exist that deal with the combinations of the two subjects and fewer still exist that take the reader through the practical implementation aspects.

This supplement introduces the fundamentals necessary to implement and apply these Soft Computing approaches to engineering problems using MATLAB. It takes the reader from the underlying theory to actual coding and implementation. Presenting the theory's implementation in code provides a more in depth understanding of the subject matter. The code is built from a bottom up framework; first introducing the pieces and then putting them together to perform more complex functions, and finally implementation examples. The MATLAB Notebook allows the embedding and evaluation of MATLAB code fragments in the Word document; thus providing a compact and comprehensive presentation of the Soft Computing techniques.

The first part of this supplement gives a very brief introduction to MATLAB including resources available on the World Wide Web. The second section of this supplement contains 17 chapters that mirror the chapters of the text. Chapters 2-13 have MATLAB implementations of the theory and discuss practical implementation issues. Although Chapters 14-17 do not give MATLAB implementations of the examples presented in the text, some references are given to support a more in depth study.

Acknowledgments

I would like to thank Distinguished Professor Robert E. Uhrig from The University of Tennessee and Professor Lefteri H. Tsoukalas from Purdue University for offering me the opportunity and encouraging me to write this supplement to their book entitled *Fuzzy and Neural Approaches in Engineering*. Also thanks for their review, comments, and suggestions.

My sincere thanks goes to Darryl Wrest of Honeywell for his time and effort during the review of this supplement. Thanks also go to Mark Buckner of Oak Ridge National Laboratory for his contributions to Sections 15.5 and 15.6.

This supplement would not have been possible without the foresight of the founders of The MathWorks in developing what I think is the most useful and productive engineering software package available. I have been a faithful user for the past seven years and look forward to the continued improvement and expansion of their base software package and application toolboxes. I have found few companies that provide such a high level of commitment to both quality and support.

About the Author

Dr. J. Wesley Hines is currently a Research Assistant Professor in the Nuclear Engineering Department at the University of Tennessee. He received the BS degree (Summa Cum Laude) in Electrical Engineering from Ohio University in 1985, both an MBA (with distinction) and a MS in Nuclear Engineering from The Ohio State University in 1992, and a Ph.D. in Nuclear Engineering from The Ohio State University in 1994. He graduated from the officers course of the Naval Nuclear Power School (with distinction) in 1986.

Dr. Hines teaches classes in Applied Artificial Intelligence to students from all departments in the engineering college. He is involved in several research projects in which he uses his experience in modeling and simulation, instrumentation and control, applied artificial intelligence, and surveillance & diagnostics in applying artificial intelligence methodologies to solve practical engineering problems. He is a member of the American Nuclear Society and IEEE professional societies and a member of Sigma Xi, Tau Beta Pi, Eta Kappa Nu, Alpha Nu Sigma, and Phi Kappa Phi honor societies.

For the five years prior to coming to the University of Tennessee, Dr. Hines was a member of The Ohio State University's Nuclear Engineering Artificial Intelligence Group. While there, he worked on several DOE and EPRI funded projects applying AI techniques to engineering problems. From 1985 to 1990 Dr. Hines served in the United States Navy as a nuclear qualified Naval Officer. He was the Assistant Nuclear Controls and Chemistry Officer for the Atlantic Submarine Force (1988 to 1990), and served as the Electrical Officer of a nuclear powered Ballistic Missile Submarine (1987 to 1988).

Software Description

This supplement comes with an IBM compatible disk that contains an install program. The install program will extract an MS Word notebook file (IBM) and several MATLAB functions, scripts and data files. The MS Word file, master.doc, is a copy of this supplement and can be opened into MS Word so that the code fragments in this document can be run and modified. Its size is over 4 megabytes, so I recommend a Pentium computer platform with at least 16 MB of RAM. It and the other files should not be duplicated or distributed without the written consent of the author.

The install program will default to the c:\matlab\toolbox\nn_fuzzy directory, but this can be changed. The program will extract 100 files and require about 5 megabytes of disk space. The contents.m file gives a brief description of the MATLAB files that were extracted into the directory. The following is a description of the files:

master.doc	This supplement was developed in MS Word 7.0 (IBM)
readme.txt	A test version of this software description section.
*.m	MATLAB script and function files (67)
*.mat	MATLAB data files (31)

INTRODUCTION TO THE MATLAB SUPPLEMENT

This supplement uses the mathematical tools of the educational version of MATLAB to demonstrate some of the important concepts presented in *Fuzzy and Neural Approaches in Engineering*, by Lefteri H. Tsoukalas and Robert E. Uhrig and being published by John Wiley & Sons. This book integrates the two technologies of fuzzy logic systems and neural networks. These two advanced information processing technologies have undergone explosive growth in the past few years. However, each field has developed independently of the other with its own nomenclature and symbology. Although there appears to be little that is common between the two fields, they are actually closely related and are being integrated in many applications. Indeed, these two technologies form the core of the discipline called SOFT COMPUTING, a name directly attributed to Lofti Zadeh. *Fuzzy and Neural Approaches in Engineering* integrates the two technologies and presents them in a clear and concise framework.

This supplement was written using the MATLAB notebook and Microsoft WORD ver. 7.0. The notebook allows MATLAB commands to be entered and evaluated while in the Word environment. This allows the document to both briefly explain the theoretical details and also show the MATLAB implementation. It allows the user to experiment with changing the MATLAB code fragments in order to gain a better understanding of the application.

This supplement contains numerous examples that demonstrate the practical implementation of neural, fuzzy, and hybrid processing techniques using MATLAB. Although MATLAB toolboxes for Fuzzy Logic [Jang and Gulley, 1995] and Neural Networks [Demuth and Beale, 1994] exist, they are not required to run the examples given in this supplement. This supplement should be considered to be a brief introduction to the MATLAB implementation of neural and fuzzy systems and the author strongly recommends the use of the Neural Networks Toolbox and the Fuzzy Logic Toobox for a more in depth study of these information processing technologies. Some of the examples in this supplement are not written in a general format and will have to be altered significantly for use to solve specific problems, other examples and m-files are extremely general and portable.

INTRODUCTION TO MATLAB

MATLAB is a technical computing environment that is published by The MathWorks. It can run on many platforms including windows based personal computers (windows, DOS, Liunix), Macintosh, Sun, DEC, VAX and Cray. Applications are transportable between the platforms.

MATLAB is the base package and was originally written as an easy interface to LINPACK, which is a state of the art package for matrix computations. MATLAB has functionality to perform or use:

- Matrix Arithmetic - add, divide, inverse, transpose, etc.
- Relational Operators - less than, not equal, etc.
- Logical operators - AND, OR, NOT, XOR
- Data Analysis - minimum, mean, covariance, etc.
- Elementary Functions - sin, acos, log, imaginary part, etc.
- Special Functions -- Bessel, Hankel, error function, etc.
- Numerical Linear Algebra - LU decomposition, etc.
- Signal Processing - FFT, inverse FFT, etc.
- Polynomials - roots, fit polynomial, divide, etc.
- Non-linear Numerical Methods - solve DE, minimize functions, etc.

MATLAB is also used for graphics and visualization in both 2-D and 3-D.

MATLAB is a language in itself and can be used at the command line or in m-files. There are two types of MATLAB M files: scripts and functions. A good reference for MATLAB programming is *Mastering MATLAB* by Duane Hanselman and Bruce Littlefield and published by Prentice Hall (http://www.prenhall.com/). These authors also wrote the user guide for the student edition of MATLAB.

1. Scripts are standard MATLAB programs that run as if they were typed into the command window.

2. Functions are compiled m-files that are stored in memory. Most MATLAB commands are functions and are accessible to the user. This allows the user to modify the functions to perform the desired functionality necessary for a specific application. MATLAB m-files may contain

- Standard programming constructs such as IF, else, break, while, etc.
- C style file I/O such as open, read, write formatted, etc.
- String manipulation commands: number to string, test for string, etc.
- Debugging commands: set breakpoint, resume, show status, etc.
- Graphical User Interfaces such as pull down menus, radio buttons, sliders, dialog boxes, mouse-button events, etc.
- On-Line help routines for all functions.

MATLAB also contains methods for external interfacing capabilities:
- For data import/export from ASCII files, binary, etc.
- To files and directories: chdir, dir, time, etc.
- To external interface libraries: C and FORTRAN callable external interface libraries.
- To dynamic linking libraries: (MEX files) allows C or FORTRAN routines to be linked directly into MATLAB at run time. This also allows access to A/D cards.
- Computational engine service library: Allows C and FORTRAN programs to call and access MATLAB routines.
- Dynamic Data Exchange: Allows MATLAB to communicate with other Windows applications.

MATLAB Toolboxes

Toolboxes are add-on packages that perform application-specific functions. MATLAB toolboxes are collections of functions that can be used from the command line, from scripts, or called from other functions. They are written in MATLAB and stored as m-files; this allows the user to modify them to meet his or her needs.

A partial listing of these toolboxes include:
- Signal Processing
- Image Processing
- Symbolic Math
- Neural Networks
- Statistics
- Spline
- Control System
- Robust Control
- Model Predictive Control
- Non-Linear Control
- System Identification
- Mu Analysis
- Optimization
- Fuzzy Logic
- Hi-Spec
- Chemometrics

SIMULINK

SIMULINK is a MATLAB toolbox which provides an environment for modeling, analyzing, and simulating linear and non-linear dynamic systems. SIMULINK provides a graphical user interface that supports click and drag of blocks that can be connected to form complex systems. SIMULINK functionality includes:

- Live displays that let you observe variables as the simulation runs.
- Linear approximations of non-linear systems can be made.
- MATLAB functions or MEX (C and FORTRAN) functions can be called.
- C code can be generated from your models.
- Output can be saved to file for later analysis.
- Systems of blocks can be combined into larger blocks to aid in program structuring.
- New blocks can be created to perform special functions such as simulating neural or fuzzy systems.
- Discrete or continuous simulations can be run.
- Seven different integration algorithms can be used depending on the system type: linear, stiff, etc.

SIMULINK's Real Time Workshop can be used for rapid prototyping, embedded real time control, real-time simulation, and stand-alone simulation. This toolbox automatically generates stand-alone C code.

User Contributed Toolboxes

Several user contributed toolboxes are available for download at the MATLAB FTP site: ftp.mathworks.com. by means of anonymous user access. Some that may be of interest are:

Genetic Algorithm Toolbox:
  A freeware toolbox developed by a MathWorks employee that will probably become a full toolbox in the future.

FISMAT Toolbox:
  A fuzzy inference system toolbox developed in Australia that incorporates several extensions to the fuzzy logic toolbox.

IFR-Fuzzy Toolbox:
  User contributed fuzzy-control toolbox.

There are also thousands of user contributed m-files on hundreds of topics ranging from the Microorbit mission analysis to sound spectrogram printing, to Lagrange interpolation. In addition to these, there are also several other MATLAB tools that are published by other companies. The most relevant of these is the Fuzzy Systems Toolbox developed by Mark H. Beale and Howard B. Demuth of the University of Idaho which is published by PWS (http://www.thomson.com/pws/default.html). This toolbox goes into greater detail than the MATLAB toolbox and better explains the lower level programming used in the functions. These authors also wrote the MATLAB Neural Network Toolbox.

MATLAB Publications

The following publications are available at The MathWorks WWW site.
   WWW Address: http://www.mathworks.com
   FTP Address: ftp.mathworks.com
      Login: anonymous
      Password: "your user address"

List of MATLAB based books.
MATLAB Digest: electronic newsletter.
MATLAB Quarterly Newsletter: News and Notes.
MATLAB Technical Notes
MATLAB Frequently Asked Questions
MATLAB Conference Archive; this conference is held every other year.
MATLAB USENET Newsgroup archive.
FTP Server also provides technical references such as papers and article reprints.

MATLAB APPLICATIONS

The following chapters implement the theory and techniques discussed in the text. These MATLAB implementations can be executed by placing the cursor in the code fragment and selecting "evaluate cell" located in the Notebook menu. The executable code fragments are green when viewed in the Word notebook and the answers are blue. Since this supplement is printed in black and white, the code fragments will be represented by 10 point Courier New gray scale. The regular text is in 12 point Times New Roman black.

Some of these implementations use m-file functions or data files. These are included on the disk that comes with this supplement. Also included is a MS Word file of this document. The file :contents.m lists and gives a brief description of all the m-files included with this supplement.

The following code segment is an autoinit cell and is executed each time the notebook is opened. If it does not execute when the document is opened, execute it manually. It performs three functions:

1. whitebg([1 1 1]) gives the figures a white background.
2. set(0, 'DefaultAxesColorOrder', [0 0 0]); close(gcf) sets the line colors in all figures to black. This produces black and white pages for printing but can be deleted for color.
3. d:/nn_fuzzy changes the current MATLAB directory to the directory where the m-files associated with this supplement are located. If you installed the files in another directory, you need to change the code to point to the directory where they are installed.

```
whitebg([1 1 1]);
set(0, 'DefaultAxesColorOrder', [0 0 0]); close(gcf)
cd d:/nn_fuzzy
```

Chapter 1 Introduction to Hybrid Artificial Intelligence Systems

Chapter 1 of *Fuzzy and Neural Approaches in Engineering*, gives a brief description of the benefits of integrating Fuzzy Logic, Neural Networks, Genetic Algorithms, and Expert Systems. Several applications are described but no specific algorithms or architectures are presented in enough detail to warrant their implementation in MATLAB.

In the following chapters, the algorithms and applications described in *Fuzzy and Neural Approaches in Engineering* will be implemented in MATLAB code. This code can be run from the WORD Notebook when in the directory containing the m-files associated with this supplement is the active directory in MATLAB's command window. In many of the chapters, the code must be executed sequentially since earlier code fragments may create data or variables used in later fragments.

Chapters 1 through 6 implement Fuzzy Logic, Chapters 7 through 11 implement Artificial Neural Networks, Chapters 12 and 13 implement fuzzy-neural hybrid systems, Chapters 14 through 17 do not contain MATLAB implementations but do point the reader towards references or user contributed toolboxes. This supplement will be updated and expanded as suggestions are received and as time permits. Updates are expected to be posted at John Wiley & Sons WWW page but may be posted at University of Tennessee web site. Further information should be available from the author at hines@utkux.utk.edu.

Chapter 2 Foundations of Fuzzy Approaches

This chapter will present the building blocks that lay the foundation for constructing fuzzy systems. These building blocks include membership functions, linguistic modifiers, and alpha cuts.

2.1 Union, Intersection and Complement of a Fuzzy Set

A graph depicting the membership of a number to a fuzzy set is called a Zadeh diagram. A Zadeh diagram is a graphical representation that shows the membership of crisp input values to fuzzy sets. The Zadeh diagrams for two membership functions A (*small numbers*) and B (*about 8*) are constructed below.

```
x=[0:0.1:20];
muA=1./(1+(x./5).^3);
muB=1./(1+.3.*(x-8).^2);
plot(x,muA,x,muB);
title('Zadeh diagram for the Fuzzy Sets A and B');
text(1,.8,'Set A');text(7,.8,'Set B')
xlabel('Number');ylabel('Membership');
```

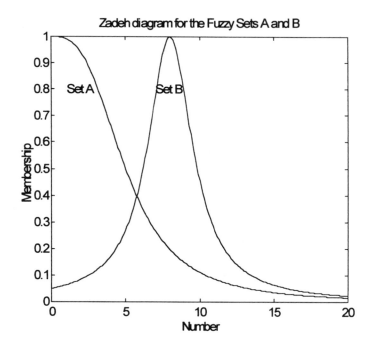

The horizontal axis of a Zadeh diagram is called the *universe of discourse*. The universe of discourse is the range of values where the fuzzy set is defined. The vertical axis is the membership of a value, in the universe of discourse, to the fuzzy set. The membership of a number (x) to a fuzzy set A is represented by: $\mu_A(x)$.

The union of the two fuzzy sets is calculated using the max function. We can see that this results in the membership of a number to the union being the maximum of its membership to either of the two initial fuzzy sets. The union of the fuzzy sets A and B is calculated below.

```
union=max(muA,muB);plot(x,union);
title('Union of the Fuzzy Sets A and B');
xlabel('Number');
ylabel('Membership');
```

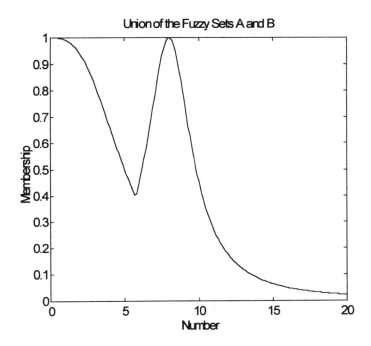

The intersection of the two fuzzy sets is calculated using the min function. We can see that this results in the membership of a number to the intersection being the minimum of its membership to either of the two initial fuzzy sets. The intersection of the fuzzy sets A and B is calculated below.

```
intersection=min(muA,muB);
plot(x,intersection);
title('Intersection of the Fuzzy Sets A and B');
xlabel('Number');
ylabel('Membership');
```

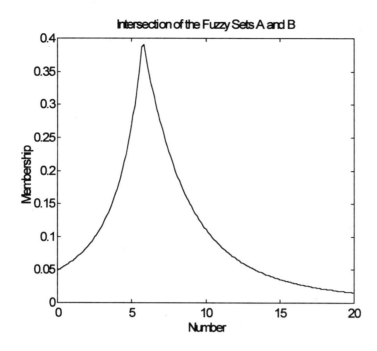

The complement of *about 8* is calculated below.

```
complement=1-muB;
plot(x,complement);
title('Complement of the Fuzzy Set B');
xlabel('Number');
ylabel('Membership');
```

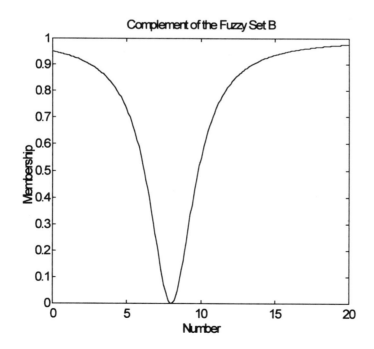

2.2 Concentration and Dilation

The concentration of a fuzzy set is equivalent to linguistically modifying it by the term *VERY*. The concentration of *small numbers* is therefore *VERY* small numbers and can be quantitatively represented by squaring the membership value. This is computed in the function very(mf).

```
x=[0:0.1:20];
muA=1./(1+(x./5).^3);
muvsb=very(muA);
plot(x,muA,x,muvsb);
title('Zadeh diagram for the Fuzzy Sets A and VERY A');
xlabel('Number');
ylabel('Membership');
text(1,.5,'Very A');
text(7,.5,'Set A')
```

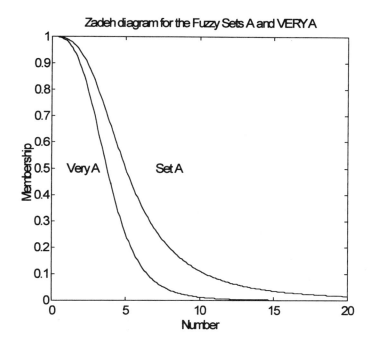

The dilation of a fuzzy set is equivalent to linguistically modifying it by the *term MORE OR LESS*. The dilation of *small numbers* is therefore *MORE OR LESS small numbers* and can be quantitatively represented by taking the square root of the membership value. This is compute in the function moreless(mf).

```
x=[0:0.1:20];
muA=1./(1+(x./5).^3);
muvsb=moreless(muA);
plot(x,muA,x,muvsb);
title('Zadeh diagram for the Fuzzy Sets A and MORE or LESS A');
xlabel('Number');
ylabel('Membership');
text(2,.5,'Set A');
```

```
text(9,.5,'More or Less A')
```

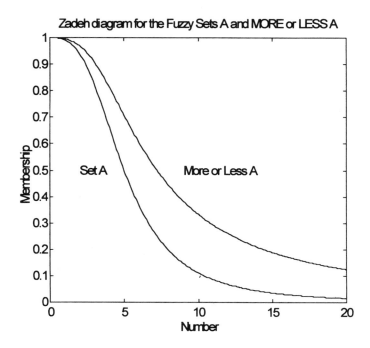

2.3 Contrast Intensification

A fuzzy set can have its fuzziness intensified, this is called contrast intensification. A membership function can be represented by an exponential fuzzifier F_1 and a denominator fuzzifier F_2. The following equation describes a fuzzy set *large numbers*.

$$\mu(x) = \frac{1}{1+\left(\dfrac{x}{F_2}\right)^{-F_1}}$$

Letting F_1 vary {1 2 4 10 100} with $F_2 = 50$ results in a family of curves with slopes increasing as F_1 increases.

```
F1=[1 2 4 10 100];
F2=50;
x=[0:1:100];
muA=zeros(length(F1),length(x));
for i=1:length(F1);
   muA(i,:)=1./(1+(x./F2).^(-F1(i)));
end
plot(x,muA);
title('Contrast Intensification');
xlabel('Number')
ylabel('Membership')
text(5,.3,'F1 = 1');text(55,.2,'F1 = 100');
```

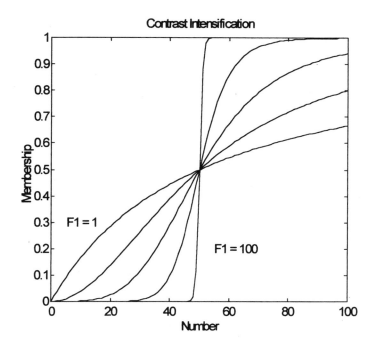

Letting F_2 vary {40 50 60 70} with $F_1 = 4$ results in the following family of curves.

```
F1=4;F2=[30 40 50 60 70];
for i=1:length(F2);
   muA(i,:)=1./(1+(x./F2(i)).^(-F1));
end
plot(x,muA);title('Contrast Intensification');
xlabel('Number');ylabel('Membership')
text(10,.5,'F2 = 30');text(75,.5,'F2 = 70');
```

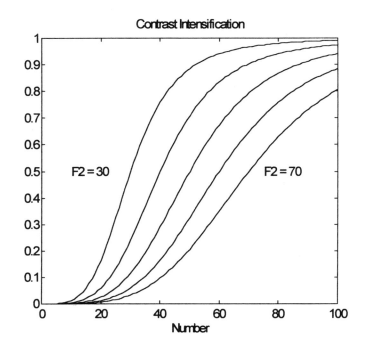

2.4 Extension Principle

The extension principle is a mathematical tool for extending crisp mathematical notions and operations to the milieu of fuzziness. Consider a function that maps points from the X-axis to the Y-axis in the Cartesian plane:

$$y = f(x) = \sqrt{1 - \frac{x^2}{4}}$$

This is graphed as the upper half of an ellipse centered at the origin.

```
x=[-2:.1:2];
y=(1-x.^2/4).^.5;
plot(x,y,x,-y);
title('Functional Mapping')
xlabel('x');ylabel('y');
```

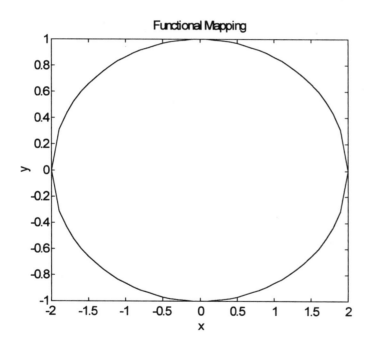

Suppose fuzzy set A is defined as $A = \int_{-2 \leq x \leq 2} \frac{\frac{1}{2}|x|}{x}$

```
mua=0.5.*abs(x);
plot(x,mua)
title('Fuzzy Set A');
xlabel('x');
ylabel('Membership of x to A');
```

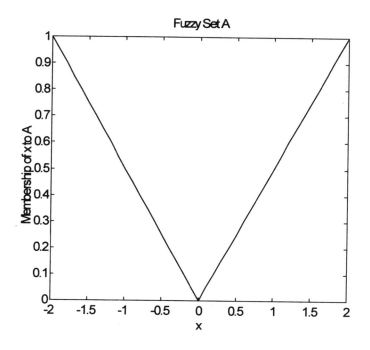

Solving for x in terms of y we get: $x = 2\sqrt{1-y^2}$.

And the membership function of y to B is $\mu_B(y) = \sqrt{1-y^2}$.

```
y=[-1:.05:1];
mub=(1-y.^2).^.5;
plot(y,mub)
title('Fuzzy Set B');
xlabel('y');ylabel('Membership of y to B');
```

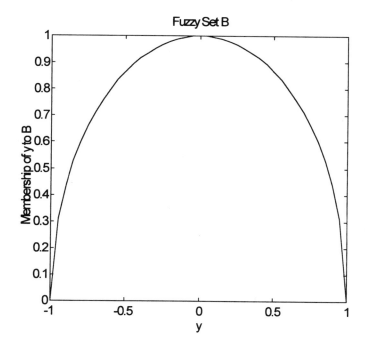

The geometric interpretation is shown below.

```
set(gcf,'color',[1 1 1]);
x=[-2:.2:2];
mua=0.5.*abs(x);
y=[-1:.1:1];
mub=(1-y.^2).^.5;
[X,Y] = meshgrid(x,y);
Z=.5*abs(X).*(1-Y.^2).^.5;
mesh(X,Y,Z);
axis([-2 2 -1 1 -1 1])
colormap(1-gray)
view([0 90]);
shading interp
xlabel('x')
ylabel('y')
title('Fuzzy Region Inside and Outside the Eclipse')
```

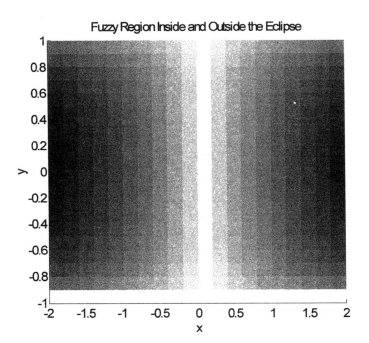

2.5 Alpha Cuts

An alpha cut is a crisp set that contains the elements that have a support (also called membership or grade) greater than a certain value. Consider a fuzzy set whose membership function is $\mu_A(x) = \frac{1}{1+0.01*(x-50).^2}$. Suppose we are interested in the portion of the membership function where the support is greater than 0.2. The 0.2 alpha cut is given by:

```
x=[0:1:100];
```

```
mua=1./(1+0.01.*(x-50).^2);
alpha_cut = mua>=.2;
plot(x,alpha_cut)
title('0.2 Level Fuzzy Set of A');
xlabel('x');
ylabel('Membership of x');
```

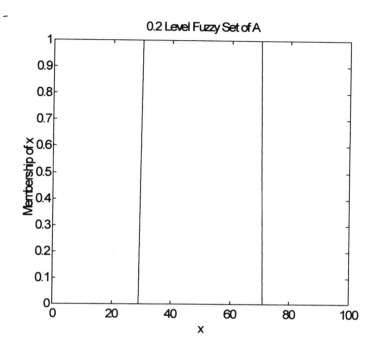

The function alpha is written to return the minimum and maximum values where an alpha cut is one. This function will be used in subsequent exercises.

```
function [a,b] = alpha(FS,x,level);
% [a,b] = alpha(FS,x,level)
%
% Returns the alpha cut for the fuzzy set at a given level.
% FS    : the grades of a fuzzy set.
% x     : the universe of discourse
% level : the level of the alpha cut
% [a,b] : the vector indices of the alpha cut
%
ind=find(FS>=level);
a=x(min(ind));
b=x(max(ind));
```

```
[a,b]=alpha(mua,x,.2)

a =
    30
b =
    70
```

Chapter 3 Fuzzy Relationships

Fuzzy *if/then* rules and their aggregations are fuzzy relations in linguistic disguise and can be thought of as fuzzy sets defined over high dimensional universes of discourse.

3.1 A Similarity Relation

Suppose a relation R is defined as "x is *near the origin AND near y*". This can be expressed as $\mu_R(x) = e^{-(x^2+y^2)}$. The universe of discourse is graphed below.

```
[x,y]=meshgrid(-2:.2:2,-2:.2:2);
mur=exp(-1*(x.^2+y.^2));
surf(x,y,mur)
xlabel('x')
ylabel('y')
zlabel('Membership to the Fuzzy Set R')
```

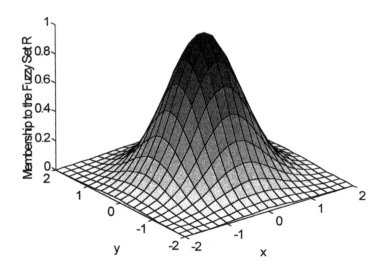

3.2 Union and Intersection of Fuzzy Relations

Suppose a relation *R1* is defined as "x is *near y AND near the origin*", and a relation *R2* is defined as "x is *NOT near the origin*". The union *R1 OR R2* is defined as:

```
mur1=exp(-1*(x.^2+y.^2));
mur2=1-exp(-1*(x.^2+y.^2));
surf(x,y,max(mur1,mur2))
xlabel('x')
ylabel('y')
zlabel('Union of R1 and R2')
```

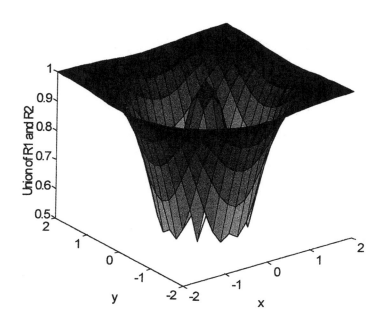

The intersection *R1 AND R2* is defined as:

```
mur1=exp(-1*(x.^2+y.^2));
mur2=1-exp(-1*(x.^2+y.^2));
surf(x,y,min(mur1,mur2))
xlabel('x')
ylabel('y')
zlabel('Intersection of R1 and R2')
```

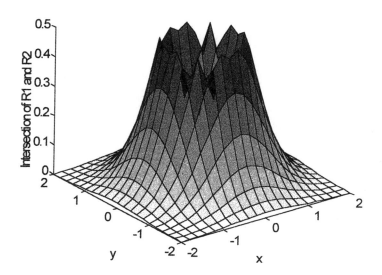

3.3 Max-Min Composition

The max-min composition uses the max and min operators described in section 3.2. Suppose two relations are defined as follows:

$$R_1 = \begin{bmatrix} \mu_{R1}(x_1,y_1) & \mu_{R1}(x_1,y_2) & \mu_{R1}(x_1,y_3) & \mu_{R1}(x_1,y_4) \\ \mu_{R1}(x_2,y_1) & \mu_{R1}(x_2,y_2) & \mu_{R1}(x_2,y_3) & \mu_{R1}(x_2,y_4) \\ \mu_{R1}(x_3,y_1) & \mu_{R1}(x_3,y_2) & \mu_{R1}(x_3,y_3) & \mu_{R1}(x_3,y_4) \\ \mu_{R1}(x_4,y_1) & \mu_{R1}(x_4,y_2) & \mu_{R1}(x_4,y_3) & \mu_{R1}(x_4,y_4) \end{bmatrix} = \begin{bmatrix} 1.0 & 0.3 & 0.9 & 0.0 \\ 0.3 & 1.0 & 0.8 & 1.0 \\ 0.9 & 0.8 & 1.0 & 0.8 \\ 0.0 & 1.0 & 0.8 & 1.0 \end{bmatrix}$$

$$R_2 = \begin{bmatrix} \mu_{R2}(x_1,y_1) & \mu_{R2}(x_1,y_2) & \mu_{R2}(x_1,y_3) \\ \mu_{R2}(x_2,y_1) & \mu_{R2}(x_2,y_2) & \mu_{R2}(x_2,y_3) \\ \mu_{R2}(x_3,y_1) & \mu_{R2}(x_3,y_2) & \mu_{R2}(x_3,y_3) \\ \mu_{R2}(x_4,y_1) & \mu_{R2}(x_4,y_2) & \mu_{R2}(x_4,y_3) \end{bmatrix} = \begin{bmatrix} 1.0 & 1.0 & 0.9 \\ 1.0 & 0.0 & 0.5 \\ 0.3 & 0.1 & 0.0 \\ 0.2 & 0.3 & 0.1 \end{bmatrix}$$

Their max-min composition is defined in its matrix form as:

$$R_1 \circ R_2 = \begin{bmatrix} 1.0 & 0.3 & 0.9 & 0.0 \\ 0.3 & 1.0 & 0.8 & 1.0 \\ 0.9 & 0.8 & 1.0 & 0.8 \\ 0.0 & 1.0 & 0.8 & 1.0 \end{bmatrix} \circ \begin{bmatrix} 1.0 & 1.0 & 0.9 \\ 1.0 & 0.0 & 0.5 \\ 0.3 & 0.1 & 0.0 \\ 0.2 & 0.3 & 0.1 \end{bmatrix}$$

Using MATLAB to compute the max-min composition:

```
R1=[1.0 0.3 0.9 0.0;0.3 1.0 0.8 1.0;0.9 0.8 1.0 0.8;0.0 1.0 0.8 1.0];
R2=[1.0 1.0 0.9;1.0 0.0 0.5; 0.3 0.1 0.0;0.2 0.3 0.1];
[r1,c1]=size(R1);   [r2,c2]=size(R2);
R0=zeros(r1,c2);
for i=1:r1;
   for j=1:c2;
      R0(i,j)=max(min(R1(i,:),R2(:,j)'));
   end
end
R0
```

```
R0 =
    1.0000    1.0000    0.9000
    1.0000    0.3000    0.5000
    0.9000    0.9000    0.9000
```

```
1.0000    0.3000    0.5000
```

Chapter 4 Fuzzy Numbers

Fuzzy numbers are fuzzy sets used in connection with applications where an explicit representation of the ambiguity and uncertainty found in numerical data is desirable.

4.1 Addition and Subtraction of Discrete Fuzzy Numbers

Addition of two fuzzy numbers can be performed using the extension principle.

Suppose you have two fuzzy numbers that are represented tabularly. They are *the fuzzy number 3* (*FN3*) and the *fuzzy number 7* (*FN7*).

FN3=0/0 + 0.3/1 + 0.7/2 + 1.0/3 + 0.7/4 + 0.3/5 + 0/6
FN7=0/4 + 0.2/5 + 0.6/6 + 1.0/7 + 0.6/8 + 0.2/9 + 0/10

To define these fuzzy numbers using MATLAB:

```
x = [1 2 3 4 5 6 7 8 9 10];
FN3 = [0.3 0.7 1.0 0.7 0.3 0 0 0 0 0];
FN7 = [0 0 0 0 0.2 0.6 1.0 0.6 0.2 0];
bar(x',[FN3' FN7']); axis([0 11 0 1.1])
title('Fuzzy Numbers 3 and 7');
xlabel('x');
ylabel('membership')
text(2,1.05,'Fuzzy Number 3')
text(6,1.05,'Fuzzy Number 7');;
```

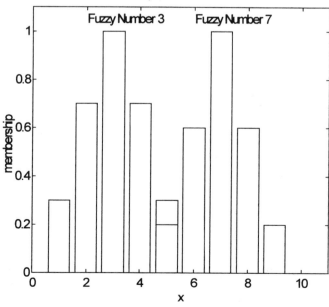

Adding *fuzzy number 3* to *fuzzy number 7* results in a *fuzzy number 10* using the alpha cut procedure described in the book.

By hand we have:
```
                 FN3      FN7      FN10 = FN3+FN7
0.2 alpha cut:   [1  5]   [5  9]   [6  14]
0.3 alpha cut:   [1  5]   [6  8]   [7  13]
0.6 alpha cut:   [2  4]   [6  8]   [8  12]
0.7 alpha cut:   [2  4]   [7  7]   [9  11]
1.0 alpha cut:   [3  3]   [7  7]   [10 10]
```

FN10 = .2/6 + .3/7 + .6/8 + .7/9 + 1/10 + .7/11 + .6/12 + .3/13 + .2/14

```
x=[1:1:20];
FNSUM=zeros(size(x));
for i=.1:.1:1
   [a1,b1]=alpha(FN3,x,i-eps);   % Use eps due to buggy MATLAB increments
   [a2,b2]=alpha(FN7,x,i-eps);
   a=a1+a2;
   b=b1+b2;
   FNSUM(a:b)=i*ones(size(FNSUM(a:b)));
end
bar(x,FNSUM); axis([0 20 0 1.1])
title('Fuzzy Number 3+7=10')
xlabel('x')
ylabel('membership')
```

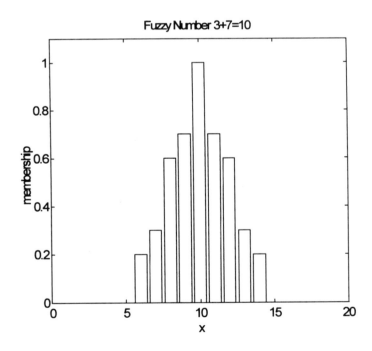

The following program subtracts the *fuzzy number 3* from the *fuzzy number 8* to get a *fuzzy number 8-3=5*.

By hand we have:

	FN3	FN8	FN5 = FN8-FN3
0.2 alpha cut:	[1 5]	[6 10]	[1 9]
0.3 alpha cut:	[1 5]	[7 9]	[2 8]
0.6 alpha cut:	[2 4]	[7 9]	[3 7]
0.7 alpha cut:	[2 4]	[8 8]	[4 6]
1.0 alpha cut:	[3 3]	[8 8]	[5 5]

FN5 = .2/1 + .3/2 + .6/3 + .7/4 + 1/5 + .7/6 + .6/7 + .3/8 + .2/9

```
x=[1:1:11];
FN3 = [0.3 0.7 1.0 0.7 0.3 0 0 0 0 0];
FN8 = [0 0 0 0 0 0.2 0.6 1.0 0.6 0.2];
FNDIFF=zeros(size(x));
for i=.1:.1:1
   [a1,a2]=alpha(FN8,x,i-eps);
   [b1,b2]=alpha(FN3,x,i-eps);
   a=a1-b2;
   b=a2-b1;
   FNDIFF(a:b)=i*ones(size(FNDIFF(a:b)));
end
bar(x,FNDIFF);axis([0 11 0 1.1])
title('Fuzzy Number 8-3=5')
xlabel('x')
ylabel('Membership')
```

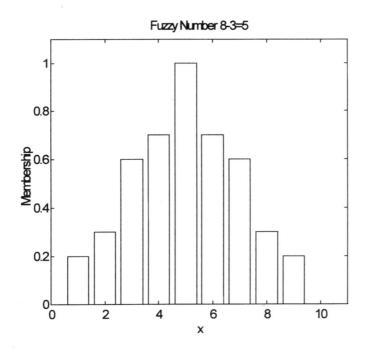

4.2 Multiplication of Discrete Fuzzy Numbers

This program multiplies the *fuzzy number 3* by the *fuzzy number 7* to get a *fuzzy number 3*7=21*. Where the fuzzy numbers *3* and *7* are defined as in Section 4.1. The multiplication of continuous fuzzy numbers is somewhat messy and will not be implemented in MATLAB.

By hand we have:
	FN3	FN7	FN21 = FN3*FN7
0.2 alpha cut:	[1 5]	[5 9]	[5 45]
0.3 alpha cut:	[1 5]	[6 8]	[6 40]
0.6 alpha cut:	[2 4]	[6 8]	[12 32]
0.7 alpha cut:	[2 4]	[7 7]	[14 28]
1.0 alpha cut:	[3 3]	[7 7]	[21 21]

FN21 = .2/5 + .3/6 + .6/12 + .7/14 + 1/21 + .7/28 + .6/32 + .3/40 + .2/45

```
x=[1:1:60];           % Universe of Discourse
FN3 = [0.3 0.7 1.0 0.7 0.3 0 0 0 0 0 0];
FN7 = [0 0 0 0 0.2 0.6 1.0 0.6 0.2 0 0];
FNPROD=zeros(size(x));
for i=.1:.1:1
   [a1,a2]=alpha(FN3,x,i-eps);
   [b1,b2]=alpha(FN7,x,i-eps);
   a=a1*b1;
   b=a2*b2;
   FNPROD(a:b)=i*ones(size(FNPROD(a:b)));
end
bar(x,FNPROD);axis([0 60 0 1.1])
title('Fuzzy Number 3*7=21')
xlabel('Fuzzy Number 21')
ylabel('Membership')
```

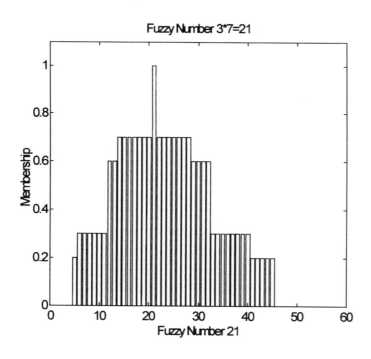

4.3 Division of Discrete Fuzzy Numbers

This program divides the *fuzzy number 6* by the *fuzzy number 3* to get a *fuzzy number 2*. The division of continuous fuzzy numbers is somewhat messy and will not be implemented in MATLAB.

By hand we have:
	FN3	FN6	FN2 = FN6/FN3
0.2 alpha cut:	[1 5]	[4 8]	[4/5 8/1]
0.3 alpha cut:	[1 5]	[5 7]	[5/5 7/1]
0.6 alpha cut:	[2 4]	[5 7]	[5/4 7/2]
0.7 alpha cut:	[2 4]	[6 6]	[6/4 6/2]
1.0 alpha cut:	[3 3]	[6 6]	[6/3 6/3]

FN21 = .2/.8 + .3/1 + .6/1.25 + .7/1.5 + 1/2 + .7/3 + .6/3.5 + .3/7 + .2/8

```
x=[1:1:12];           % Universe of Discourse
FN3 = [0.3 0.7 1.0 0.7 0.3 0 0 0 0 0];
FN6 = [0 0 0 0.2 0.6 1.0 0.6 0.2 0 0];
FNDIV=zeros(size(x));
for i=.1:.1:1
   [a1,a2]=alpha(FN6,x,i-eps);
   [b1,b2]=alpha(FN3,x,i-eps);
   a=round(a1/b2);
   b=round(a2/b1);
   FNDIV(a:b)=i*ones(size(FNDIV(a:b)));
end
bar(x,FNDIV);axis([0 10 0 1.1])
title('Fuzzy Number 6/3=2')
xlabel('Fuzzy Number 2')
ylabel('Membership')
```

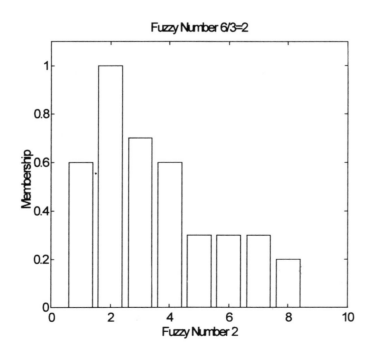

Chapter 5 Linguistic Descriptions and Their Analytical Form

5.1 Generalized Modus Ponens

Fuzzy linguistic descriptions are formal representations of systems made through fuzzy *if/then* rules. Generalized Modus Ponens (GMP) states that when a rule's antecedent is met to some degree, its consequence is inferred by the same degree.

IF   x is *A*   THEN   y is *B*
         x is *A'*
so   y is *B'*

This can be written using the implication relation ($R(x,y)$) as in the max-min composition of section 3.3.

$B' = A' \circ R(x,y)$

Implication relations are explained in greater detail in section 5.3.

5.2 Membership Functions

This supplement contains functions that define triangular, trapezoidal, S-Shaped and Π-shaped membership functions.

5.2.1 Triangular Membership Function

A triangular membership function is defined by the parameters [a b c], where a is the membership function's left intercept with grade equal to 0, b is the center peak where the grade equals 1 and c is the right intercept at grade equal to 0. The function y=triangle(x,[a b c]); is written to return the membership values corresponding to the defined universe of discourse x. The parameters that define the triangular membership function: [a b c] must be in the discretely defined universe of discourse.

For example: A triangular membership function for "*x is close to 33*" defined over x=[0:1:50] with [a b c]=[23 33 43] would be created with:

```
x=[0:1:50];
y=triangle(x,[23 33 43]);
plot(x,y);
title('Close to 33')
xlabel('X')
ylabel('Membership')
```

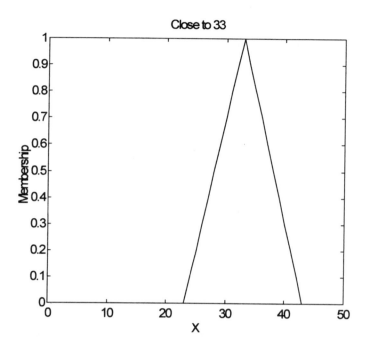

A fuzzy variable temperature may have three fuzzy values: *cool, medium* and *hot*. Membership functions defining these values can be constructed to overlap in the universe of discourse [0:100]. A matrix with each row corresponding to the three fuzzy values can be constructed. Suppose the following fuzzy value definitions are used:

```
x=[0:100];
cool=[0 25 50];
medium=[25 50 75];
hot=[50 75 100];
mf_cool=triangle(x,cool);
mf_medium =triangle(x,medium);
mf_hot=triangle(x,hot);
plot(x,[mf_cool;mf_medium;mf_hot])
title('Temperature: cool, medium and hot');
ylabel('Membership');
xlabel('Degrees')
text(20,.58,'Cool')
text(42,.58,'Medium')
text(70,.58,'Hot')
```

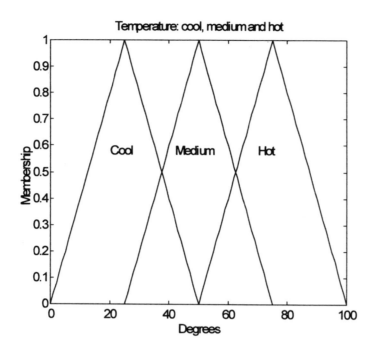

5.2.2 Trapezoidal Membership Function

As can be seen, a temperature value of 0 would have a 0 membership to all fuzzy sets. Therefore, we should use trapezoidal membership functions to define the *cool* and *hot* fuzzy sets.

```
x=[0:100];
cool=[0 0 25 50];
medium=[15 50 75];
hot=[50 75 100 100];
mf_cool=trapzoid(x,cool);
mf_medium =triangle(x,medium);
mf_hot=trapzoid(x,hot);
plot(x,[mf_cool;mf_medium;mf_hot]);
title('Temperature: cool, medium and hot');
ylabel('Membership');
xlabel('Degrees');
text(20,.65,'Cool')
text(42,.65,'Medium')
text(70,.65,'Hot')
```

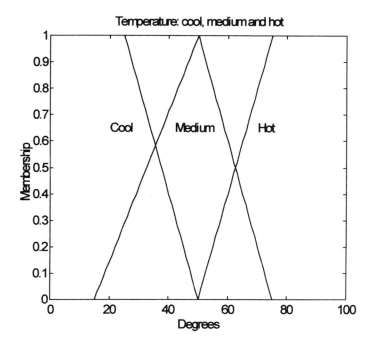

The use of trapezoidal membership functions results in a 0 value of temperature being properly represented by a membership value of 1 to the fuzzy set *cool*. Likewise, high temperatures are properly represented with high membership values to the fuzzy set *hot*.

5.2.3 S-shaped Membership Function

An S-shaped membership function is defined by three parameters [α β γ] using the following equations:

$$S\_shape(\alpha, \beta, \gamma) = 0 \qquad \text{for} \qquad x \leq \alpha$$

$$S\_shape(\alpha, \beta, \gamma) = 2\left(\frac{x-\alpha}{\gamma-\alpha}\right)^2 \qquad \text{for} \qquad \alpha \leq x \leq \beta$$

$$S\_shape(\alpha, \beta, \gamma) = 1 - 2\left(\frac{x-\gamma}{\gamma-\alpha}\right)^2 \qquad \text{for} \qquad \beta \leq x \leq \gamma$$

$$S\_shape(\alpha, \beta, \gamma) = 1 \qquad \text{for} \qquad \gamma \leq x$$

where:
  α = the point where μ(x)=0
  β = the point where μ(x)=0.5
  γ = the point where μ(x)=1.0
note: β-α must equal γ-β for continuity of slope

```
x=[0:100];
cool=[50 25 0];
hot=[50 75 100];
mf_cool=s_shape(x,cool);
mf_hot=s_shape(x,hot);
plot(x,[mf_cool;mf_hot]);
title('Temperature: cool and hot');
ylabel('Membership');
xlabel('Degrees');
text(8,.45,'Cool')
text(82,.45,'Hot')
```

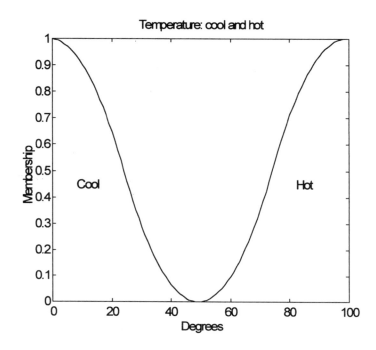

5.2.4 Π-shaped Membership Function

A Π-shaped membership functions is defined by two parameters [γ,β] using the following equations:

$$P\_shape(\gamma,\delta) = S\_shape\left(x;\gamma-\delta,\frac{\gamma-\delta}{2},\gamma\right) \quad \text{for} \quad x \leq \gamma$$

$$P\_shape(\gamma,\delta) = 1 - S\_shape\left(x;\gamma,\frac{\gamma+\delta}{2},\gamma+\delta\right) \quad \text{for} \quad x \geq \gamma$$

where:
  γ = center of the membership function
  β = width of the membership function at grade = 0.5.

```
x=[0:100];
cool=[25 20];
medium=[50 20];
hot=[75 20];
mf_cool=p_shape(x,cool);
mf_medium =p_shape(x,medium);
mf_hot=p_shape(x,hot);
plot(x,[mf_cool;mf_medium;mf_hot]);
title('Temperature: cool, medium and hot');
ylabel('Membership');
xlabel('Degrees');
text(20,.55,'Cool')
text(42,.55,'Medium')
text(70,.55,'Hot')
```

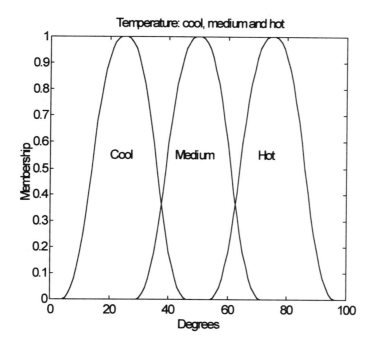

5.2.5 Defuzzification of a Fuzzy Set

Defuzzification is the process of representing a fuzzy set with a crisp number and is discussed in Section 6.3 of the text. Internal representations of data in a fuzzy system are usually fuzzy sets but the output frequently needs to be a crisp number that can be used to perform a function, such as commanding a valve to a desired position.

The most commonly used defuzzification method is the center of area method also commonly referred to as the centroid method. This method determines the center of area of the fuzzy set and returns the corresponding crisp value. The function centroid (universe, grades) performs this function by using a method similar to that of finding a balance point on a loaded beam.

```
function [center] = centroid(x,y);
%CENTER Calculates Centroid
% [center] = centroid(universe,grades)
%
% universe: row vector defining the universe of discourse.
% grades:   row vector of corresponding membership.
% centroid: crisp number defining the centroid.
%
center=(x*y')/sum(y);
```

To illustrate this method, we will defuzzify the following triangular fuzzy set and plot the result using c_plot:

```
x=[10:150];
y=triangle(x,[32 67 130]);
center=centroid(x,y);
c_plot(x,y,center,'Centroid')
```

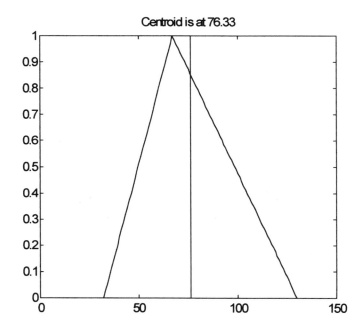

There are several other defuzzification methods including mean of max, max of max and min of max. The following function implements mean of max defuzzification: mom(universe,grades).

```
x=[10:150];
y=trapzoid(x,[32 67 74 130]);
center=mom(x,y);
c_plot(x,y,center,'Mean of Max');
```

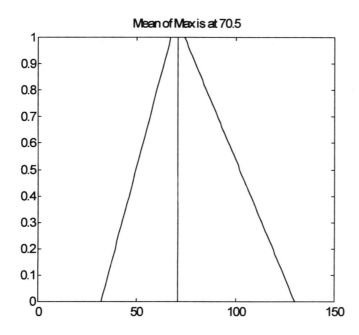

5.2.6 Compound Values

Connectives such as *AND* and *OR*, and modifiers such as *NOT*, *VERY*, and *MORE or LESS* can be used to generate compound values from primary values:

OR corresponds to max or union
AND corresponds to min or intersection
NOT corresponds to the complement and is calculated by the function not(MF).
VERY, *MORE or LESS*, etc. correspond to various degrees of contrast intensification.

Temperature is *NOT cool AND NOT hot* is a fuzzy set represented by:

```
x=[0:100];
cool=[0 0 25 50];
hot=[50 75 100 100];
mf_cool=trapzoid(x,cool);
mf_hot=trapzoid(x,hot);
not_cool=not(mf_cool);
not_hot=not(mf_hot);
answer=min([not_hot;not_cool]);
plot(x,answer);
title('Temperature is NOT hot AND NOT cool');
ylabel('Membership');
xlabel('Degrees');
```

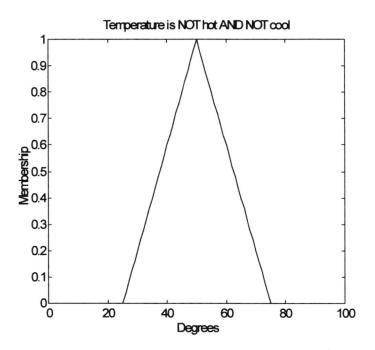

VERY and *MORE or LESS* are called linguistic modifiers. These can be implemented by taking the square (*VERY*) or square root (*MORE or LESS*) of the membership values. These modifiers are implemented with the very(MF) and moreless(MF) functions. For example, *NOT VERY hot* would be represented as:

```
not_very_hot=not(very(trapzoid(x,hot)));
plot(x,not_very_hot);
title('NOT VERY hot');ylabel('Membership');xlabel('Degrees');
```

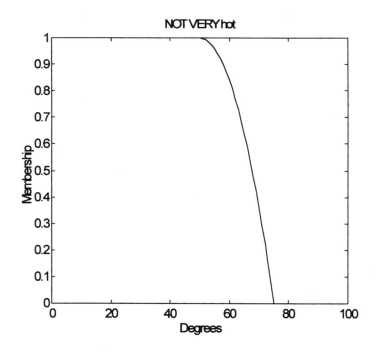

and, *MORE or LESS hot* would be represented as:

```
ml_hot=moreless(trapzoid(x,hot));
plot(x,ml_hot);
title('Temperature is More or Less hot');
ylabel('Membership');xlabel('Degrees');
```

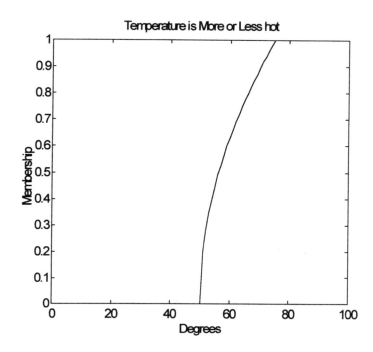

Note that some membership functions are affected by linguistic modifiers more than others. For example, a membership function that only has crisp values, such as a hardlimit membership function, would not be affected at all.

5.3 Implication Relations

The underlying analytical form of an *if/then* rule is a fuzzy relation called an implication relation: *R(x,y)*. There are several implication relation operators (ϕ) including:

Zadeh Max-Min Implication Operator $\phi[\mu_A(x), \mu_B(y)] = (\mu_A(x) \wedge \mu_B(y)) \vee (1 - \mu_A(x))$

Mamdami Min Implication Operator $\phi[\mu_A(x), \mu_B(y)] = \mu_A(x) \wedge \mu_B(y)$

Larson Product Implication Operator $\phi[\mu_A(x), \mu_B(y)] = \mu_A(x) \cdot \mu_B(y)$

To illustrate the Mamdami Min implementation operator, suppose there is a rule that states:

if   x is "*Fuzzy Number 3*"
then   y is "*Fuzzy Number 7*"

For the *Fuzzy Number 3* of section 4.1, if the input x is a 2, it matches the set "*Fuzzy Number 3*" with a value of 0.7. This value is called the "Degree of Fulfillment" (DOF) of the antecedent. Therefore, the consequence should be met with a degree of 0.7 and results in the output fuzzy number being clipped to a maximum of 0.7. To perform this operation we construct a function called clip(FS,level).

```
mua=1./(1+0.01.*(x-50).^2);
clip_mua=clip(mua,0.2);
plot(x,clip_mua);
title('Fuzzy Set A Clipped to a 0.2 Level');
xlabel('x');
ylabel('Membership of x');
```

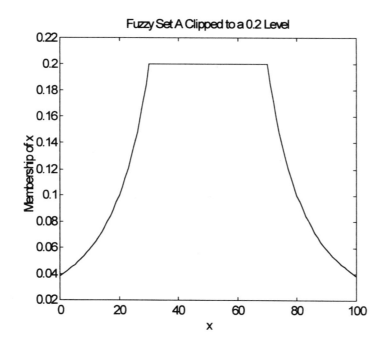

Referring back to the discrete example:
if   x is "*Fuzzy Number 3*"
then   y is "*Fuzzy number 7*"
and x is equal to 2, then the output y is equal to the fuzzy set clipped at 2's degree of fulfillment of *Fuzzy Number 7*.

```
x= [0 1 2 3 4 5 6 7 8 9 10];
FN3 = [0 0.3 0.7 1.0 0.7 0.3 0 0 0 0 0];
FN7 = [0 0 0 0 0 0.2 0.6 1.0 0.6 0.2 0];
degree=FN3(find(x==2));
y=clip(FN7,degree);
plot(x,y);
axis([0 10 0 1])
```

```
title('Mamdani Min Output of Fuzzy Rule');
xlabel('x');
ylabel('Output Fuzzy Set');
```

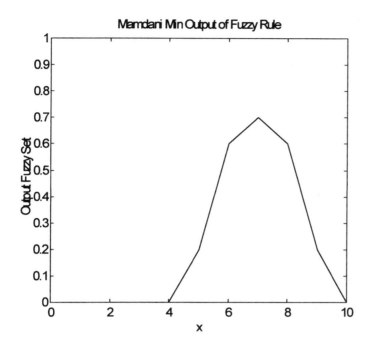

This example shows the basic foundation of a rule based fuzzy logic system. We can see that using discrete membership functions of very rough granularity may not provide the precision that one may desire. Membership functions with less granularity should be used.

To illustrate the use of the Larson Product implication relation, suppose there is a rule that states:

if   x is "*Fuzzy Number 3*"
then   y is "*Fuzzy number 7*"

For the *Fuzzy Number 3* of section 4.1, if the input x is a 2, it matches the antecedent fuzzy set "*Fuzzy Number 3*" with a degree of fulfillment of 0.7. The Larson Product implication operator scales the consequence with the degree of fulfillment which is 0.7 and results in the output fuzzy number being scaled to a maximum of 0.7. The function product(FS,level) performs the Larson Product operation.

```
x=[0:1:100];
mua=1./(1+0.01.*(x-50).^2);
prod_mua=product(mua,.7);
plot(x,prod_mua)
axis([min(x) max(x) 0 1]);
title('Fuzzy Set A Scaled to a 0.7 Level');
xlabel('x');
ylabel('Membership of x');
```

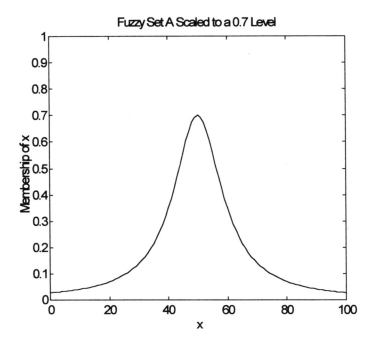

Referring back to the highly granular discrete example:

if   x is *"Fuzzy Number 3"*
then   y is *"Fuzzy Number 7"*

and x is equal to 2, then the output y is equal to the fuzzy set squashed to the antecedent's degree of fulfillment to *"Fuzzy Number 7"*.

```
x= [0 1 2 3 4 5 6 7 8 9 10];
FN3 = [0 0.3 0.7 1.0 0.7 0.3 0 0 0 0 0];
FN7 = [0 0 0 0 0 0.2 0.6 1.0 0.6 0.2 0];
degree=FN3(find(x==2));
y=product(FN7,degree);
plot(x,y);
axis([0 10 0 1.0])
title('Larson Product Output of Fuzzy Rule');
xlabel('x');
ylabel('Output Fuzzy Set');
```

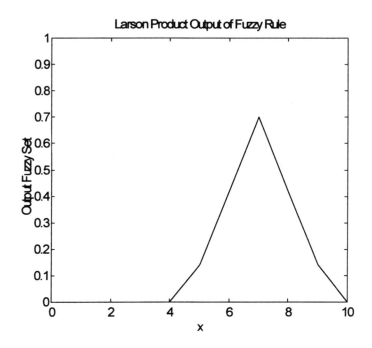

5.4 Fuzzy Algorithms

Now that we can manipulate Fuzzy Rules we can combine them into Fuzzy Algorithms. A Fuzzy Algorithm is a procedure for performing a task formulated by a collection of fuzzy *if/then* rules. These rules are usually connected by *ELSE* statements.

if	x is A_1	*then*	y is B_1	*ELSE*
if	x is A_2	*then*	y is B_2	*ELSE*
...				
if	x is A_n	*then*	y is B_n	

ELSE is interpreted differently for different implication operators:

Zadeh Max-Min Implication Operator	*AND*
Mamdami Min Implication Operator	*OR*
Larson Product Implication Operator	*OR*

As a first example, consider a fuzzy algorithm that controls a fan's speed. The input is the crisp value of temperature and the output is a crisp value for the fan speed. Suppose the fuzzy system is defined as:

if	Temperature is *Cool*	*then*	Fan_speed is *Low*	*ELSE*
if	Temperature is *Moderate*	*then*	Fan_speed is *Medium*	*ELSE*
if	Temperature is *Hot*	*then*	Fan_speed is *High*	

This system has three fuzzy rules where the antecedent membership functions *Cool*, *Moderate*, *Hot* and consequent membership functions *Low*, *Medium*, *High* are defined by the following fuzzy sets over the given universes of discourse:

```
% Universe of Discourse
x = [0:1:120];    % Temperature
y = [0:1:10];     % Fan Speed

% Temperature
cool_mf = trapzoid(x,[0 0 30 50]);
moderate_mf = triangle(x,[30 55 80]);
hot_mf = trapzoid(x,[60 80 120 120]);
antecedent_mf = [cool_mf;moderate_mf;hot_mf];
plot(x,antecedent_mf)
title('Cool, Moderate and Hot Temperatures')
xlabel('Temperature')
ylabel('Membership')
```

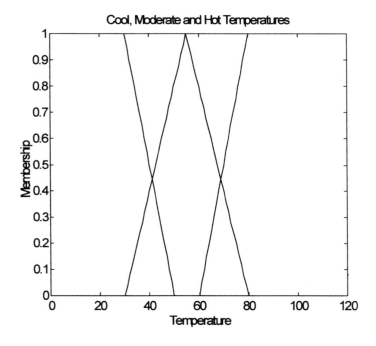

```
% Fan Speed
low_mf = trapzoid(y,[0 0 2 5]);
medium_mf = trapzoid(y,[2 4 6 8]);
high_mf = trapzoid(y,[5 8 10 10]);
consequent_mf = [low_mf;medium_mf;high_mf];
plot(y,consequent_mf)
title('Low, Medium and High Fan Speeds')
xlabel('Fan Speed')
ylabel('Membership')
```

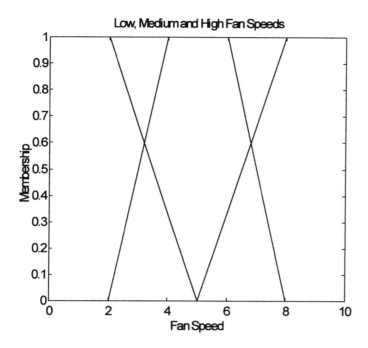

Now that we have the membership functions defined we can perform the five steps of evaluating fuzzy algorithms:

1. Fuzzify the input.
2. Apply a fuzzy operator.
3. Apply an implication operation.
4. Aggregate the outputs.
5. Defuzzify the output.

First we fuzzify the input. The output of the first step is the degree of fulfillment of each rule. Suppose the input is Temperature = 72.

```
temp = 72;
dof1 = cool_mf(find(x==temp));
dof2 = moderate_mf(find(x == temp));
dof3 = hot_mf(find(x == temp));
DOF = [dof1;dof2;dof3]
```

```
DOF =
         0
    0.3200
    0.6000
```

Doing this in matrix notation:

```
temp=72;
DOF=antecedent_mf(:,find(x==temp))
```

```
DOF =
```

39

```
    0
0.3200
0.6000
```

There is no fuzzy operator (*AND*, *OR*) since each rule has only one input. Next we apply a fuzzy implication operation. Suppose we choose the Larson Product implication operation.

```
consequent1 = product(low_mf,dof1);
consequent3 = product(medium_mf,dof2);
consequent2 = product(high_mf,dof3);
plot(y,[consequent1;consequent2;consequent3])
axis([0 10 0 1.0])
title('Consequent Fuzzy Set')
xlabel('Fan Speed')
ylabel('Membership')
```

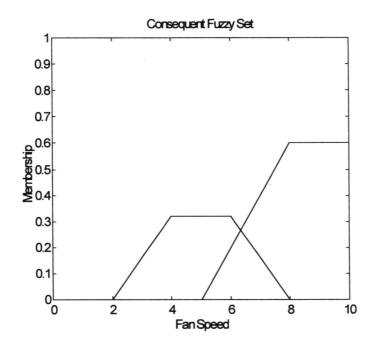

Or again, in matrix notation:

```
consequent = product(consequent_mf,DOF);
plot(y,consequent)
axis([0 10 0 1.0])
title('Consequent Fuzzy Set')
xlabel('Fan Speed')
ylabel('Membership')
```

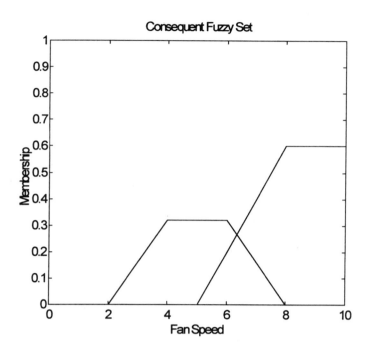

Next we need to aggregate the consequent fuzzy sets. We will use the max operator.

```
Output_mf=max([consequent1;consequent2;consequent3]);
plot(y,Output_mf)
axis([0 10 0 1])
title('Output Fuzzy Set')
xlabel('Fan Speed')
ylabel('Membership')
```

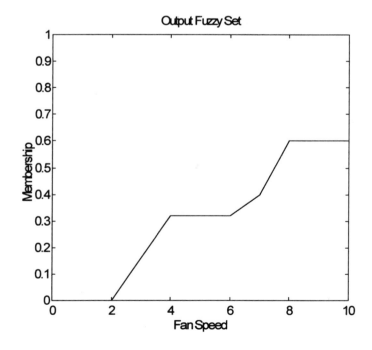

```
Output_mf = max(consequent);
plot(y,Output_mf)
axis([0 10 0 1]);title('Output Fuzzy Set')
xlabel('Fan Speed');ylabel('Membership')
```

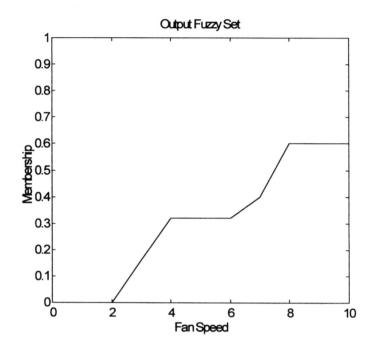

Lastly we defuzzify the output set to obtain a crisp value.

```
output=centroid(y,Output_mf);
c_plot(y,Output_mf,output,'Crisp Output');
```

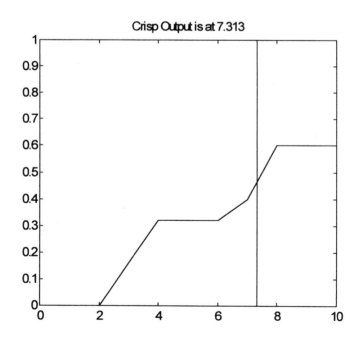

The crisp output of the fuzzy rules states that the fan speed should be set to a value of 7.3 for a temperature of 72 degrees. To see the output for different input temperatures, we write a loop that covers the input universe of discourse and computes the output for each input temperature. Note: you must have already run the code fragments that set up the membership functions and define the universe of discourse to run this example.

```
outputs=zeros(size([1:1:100]));
for temp=1:1:100
   DOF=antecedent_mf(:,find(x==temp));        %Fuzzification
   consequent = product(consequent_mf,DOF);   %Implication
   Output_mf = max(consequent);               %Aggregation
   output=centroid(y,Output_mf);              %Defuzzification
   outputs(temp)=output;
end
plot([1:1:100],outputs)
title('Fuzzy System Input Output Relationship')
xlabel('Temperature')
ylabel('Fan Speed')
```

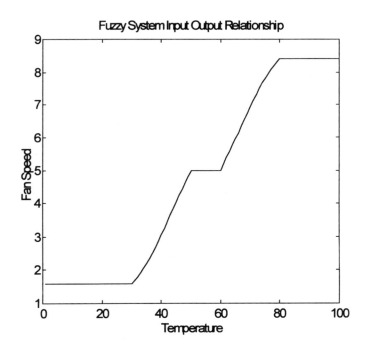

We see that the input/output relationship is non-linear. The next chapter will demonstrate fuzzy tank level control when Fuzzy Operators are included.

Chapter 6 Fuzzy Control

Fuzzy control refers to the control of processes through the use of fuzzy linguistic descriptions. For additional reading on fuzzy control see DeSilva, 1995; Jamshidi, Vadiee and Ross, 1993; or Kandel and Langholz, 1994.

6.1 Tank Level Fuzzy Control

A tank is filled by means of a valve and continuously drains. The level is measured and compared to a level setpoint forming a level error. This error is used by a controller to position the valve to make the measured level equal to the desired level. The setup is shown below and is used in a laboratory at The University of Tennessee for fuzzy and neural control experiments.

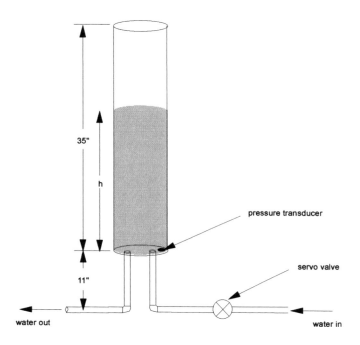

This is a nonlinear control problem since the dynamics of the plant are dependent on the height of level of the water through the square root of the level. There also may be some non-linearities due to the valve flow characteristics. The following equations model the process.

$$\dot{h} = \frac{Vin - Vout}{Area} \qquad Area = pi * R^2 = A_k$$

$$Vout = K\sqrt{h} \qquad K \text{ is the resistance in the outlet piping}$$

$$Vin = f(u) \qquad u \text{ is the valve position}$$

$$\dot{h} = \frac{f(u) - K\sqrt{h}}{A_k} = \frac{f(u)}{A_k} - \frac{K\sqrt{h}}{A_k}$$

$$\dot{h}A_k + K\sqrt{h} = f(u)$$

These equations can be used to model the plant in SIMULINK.

WATER TANK MODEL REPRESENTATION

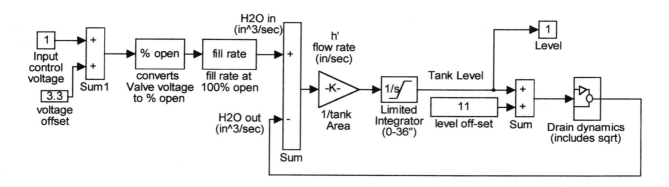

The non-linearities are apparent when linearizing the plant around different operating levels. This can be done using LINMOD.

[a,b,c,d]=linmod('tank',3,.1732051)
resulting in:
a = -0.0289   b =   1.0   c =   1.0

For different operating levels we have:
for h=1   pole at -0.05
for h=2   pole at -0.0354
for h=3   pole at -0.0289
for h=4   pole at -0.025

This nonlinearity makes control with a PID controller difficult unless gain scheduling is used. A controller designed to meet certain performance specifications at a low level such as h=1 may not meet those specifications at a higher level such as h=4. Therefore, a fuzzy controller may be a viable alternative.

The fuzzy controller described in the book uses two input variables [error, change in error] to control valve position. The membership functions were chosen to be:

Error: *nb, nm, z, pm, pb*
Change in Error: *ps, pm, pb*
Valve Position: *vh, high, med, low, vl*

Where:
nb, nm, z, pm, pb = negative big, negative medium, zero, positive big, positive medium
ps, pm, pb = positive small, positive medium, positive big
vh, high, med, low, vl = very high, high, medium, low, very low

Fifteen fuzzy rules are used to account for each combination of input variables:
1. *if* (error is *nb*) *AND* (del_error is *n*) *then* (control is *high*) (1) ELSE
2. *if* (error is *nb*) *AND* (del_error is *ze*) *then* (control is *vh*) (1) ELSE
3. *if* (error is *nb*) *AND* (del_error is *p*) *then* (control is *vh*) (1) ELSE
4. *if* (error is *ns*) *AND* (del_error is *n*) *then* (control is *high*) (1) ELSE
5. *if* (error is *ns*) *AND* (del_error is *ze*) *then* (control is *high*) (1) ELSE
6. *if* (error is *ns*) *AND* (del_error is *p*) *then* (control is *med*) (1) ELSE
7. *if* (error is *z*) *AND* (del_error is *n*) *then* (control is *med*) (1) ELSE
8. *if* (error is *z*) *AND* (del_error is *ze*) *then* (control is *med*) (1) ELSE
9. *if* (error is *z*) *AND* (del_error is *p*) *then* (control is *med*) (1) ELSE
10. *if* (error is *ps*) *AND* (del_error is *n*) *then* (control is *med*) (1) ELSE
11. *if* (error is *ps*) *AND* (del_error is *ze*) *then* (control is *low*) (1) ELSE
12. *if* (error is *ps*) *AND* (del_error is *p*) *then* (control is *low*) (1) ELSE
13. *if* (error is *pb*) *AND* (del_error is *n*) *then* (control is *low*) (1) ELSE
14. *if* (error is *pb*) *AND* (del_error is *ze*) *then* (control is *vl*) (1) ELSE
15. *if* (error is *pb*) *AND* (del_error is *p*) *then* (control is *vl*) (1)

The membership functions were manually tuned by trial and error to give good controller performance. Automatic adaptation of membership functions will be discussed in Chapter 13. The resulting membership functions are:

```
Level_error =   [-36:0.1:36];
nb = trapzoid(Level_error,[-36 -36 -10 -5]);
ns = triangle(Level_error,[-10 -2 0]);
z = triangle(Level_error,[-1 0 1]);
ps = triangle(Level_error,[0 2 10]);
pb = trapzoid(Level_error,[5 10 36 36]);
l_error = [nb;ns;z;ps;pb];
plot(Level_error,l_error);
title('Level Error Membership Functions')
xlabel('Level Error')
ylabel('Membership')
```

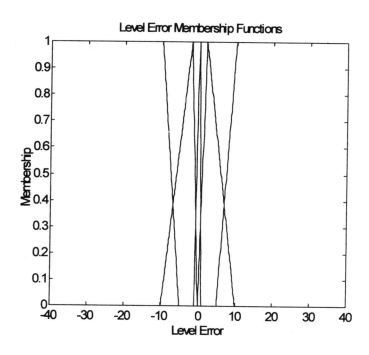

```
Del_error = [-40:.1:40];
p = trapzoid(Del_error,[-40 -40 -2 0]);
ze = triangle(Del_error,[-1 0 1]);
n = trapzoid(Del_error,[0 2 40 40]);
d_error = [p;ze;n];
plot(Del_error,d_error);
title('Level Rate Membership Functions')
xlabel('Level Rate')
ylabel('Membership')
```

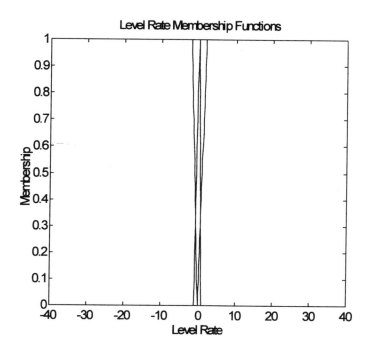

```
Control = [-4.5:0.05:1];
vh = triangle(Control,[0 1 1]);
high = triangle(Control,[-1 0 1]);
med = triangle(Control,[-3 -2 -1]);
low = triangle(Control,[-4.5 -3.95 -3]);
vl = triangle(Control,[-4.5 -4.5 -3.95]);
control=[vh;high;med;low;vl];
plot(Control,control);
title('Output Voltage Membership Functions')
xlabel('Control Voltage')
ylabel('Membership')
```

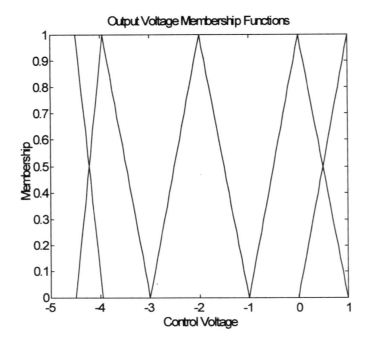

A Mamdami fuzzy system that uses centroid defuzzification will now be created. Test results show that the fuzzy system performs superior to that of a PID controller. There was practically no overshoot., and the speed of response was only limited by the inlet supply pressure and output piping resistance. Suppose the following error and change in error are input to the fuzzy controller. First, the degree of fulfillments of the antecedent membership functions are calculated.

```
error=-8.1;
derror=0.3;
DOF1=interp1(Level_error',l_error',error')';
DOF2=interp1(Del_error',d_error',derror')';
```

Next, the fuzzy relation operations inherent in the 15 rules are performed.

```
antecedent_DOF = [min(DOF1(1), DOF2(1))
 min(DOF1(1), DOF2(2))
 min(DOF1(1), DOF2(3))
```

```
        min(DOF1(2), DOF2(1))
        min(DOF1(2), DOF2(2))
        min(DOF1(2), DOF2(3))
        min(DOF1(3), DOF2(1))
        min(DOF1(3), DOF2(2))
        min(DOF1(3), DOF2(3))
        min(DOF1(4), DOF2(1))
        min(DOF1(4), DOF2(2))
        min(DOF1(4), DOF2(3))
        min(DOF1(5), DOF2(1))
        min(DOF1(5), DOF2(2))
        min(DOF1(5), DOF2(3))]

antecedent_DOF =
        0
   0.6200
   0.1500
        0
   0.2375
   0.1500
        0
        0
        0
        0
        0
        0
        0
        0
        0

consequent = [control(5,:)
control(5,:)
control(4,:)
control(4,:)
control(4,:)
control(3,:)
control(3,:)
control(3,:)
control(3,:)
control(3,:)
control(2,:)
control(2,:)
control(2,:)
control(1,:)
control(1,:)];

Consequent = product(consequent,antecedent_DOF);
plot(Control,Consequent)
axis([min(Control) max(Control) 0 1.0])
title('Consequent of Fuzzy Rules')
xlabel('Control Voltage')
ylabel('Membership')
```

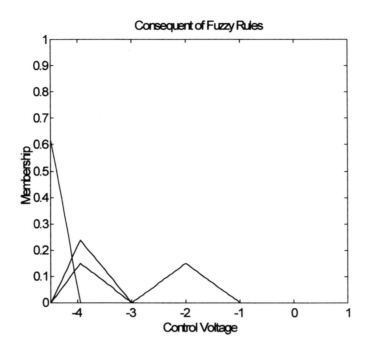

The fuzzy output sets are aggregated to form a single fuzzy output set.

```
aggregation = max(Consequent);
plot(Control,aggregation)
axis([min(Control) max(Control) 0 1.0])
title('Aggregation of Fuzzy Rule Outputs')
xlabel('Control Voltage')
ylabel('Membership')
```

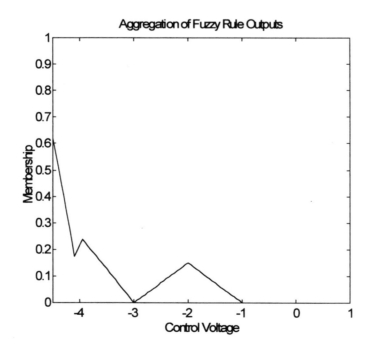

The output fuzzy set is defuzzified to find the crisp output voltage.

```
output=centroid(Control,aggregation);
c_plot(Control,aggregation,output,'Crisp Output Value for Voltage')
axis([min(Control) max(Control) 0 1.0])
xlabel('Control Voltage');
```

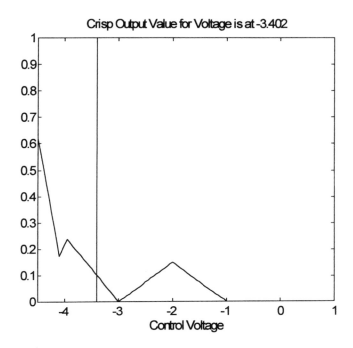

For these inputs, a voltage of -3.4 would be sent to the control valve.

Now that we have the five steps of evaluating fuzzy algorithms defined (fuzzification, apply fuzzy operator, apply implication operation, aggregation and defuzzification), we can combine them into a function that is called at each controller voltage update. The level error and change in level error will be passed to the fuzzy controller function and the command valve actuator voltage will be passed back. This function, named tankctrl(), is included as an m-file. The universes of discourse and membership functions are initialized by a MATLAB script named tankinit. These variables are made to be global MATLAB variables because they need to be used by the fuzzy controller function.

The differential equations that model the tank are contained in a function called tank_mod.m. It operates by passing to it the current state of the tank (tank level) and the control valve voltage. It passes back the next state of the tank. A demonstration of the operation of the tank with its controller is given in the function tankdemo(initial_level,desired_level). You may try running tankdemo with different initial and target levels. This function plots out the result of a 40 second simulation, this may take from 10 seconds to a minute or two depending on the speed of the computer used for the simulation.

`tankdemo(24.3,11.2)`

```
The tank and controller are simulated for 40 seconds, please be patient.
```

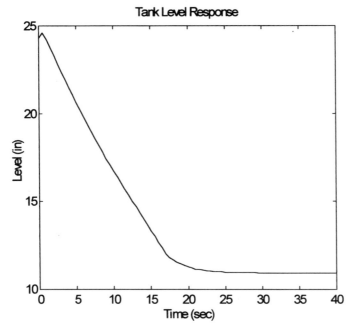

As you can see, the controller has very good response characteristics. There is very low steady state error and no overshoot. The speed of response is mostly controlled by the piping and valve resistances. The first second of the simulation is before feedback occurs, so disregard that data point.

By changing the membership functions and rules, you can get different response characteristics. The steady state error is controlled by the width of the zero level error membership function. Keeping this membership function thin, keeps the steady state error small.

Chapter 7 Fundamentals of Neural Networks

The MathWorks markets a Neural Networks Toolbox. A description of it can be found at http://www.mathworks.com/neural.html. Other MATLAB based Neural Network tools are the NNSYSID Toolbox at http://kalman.iau.dtu.dk/Projects/proj/nnsysid.html and the NNCTRL toolkit at http://www.iau.dtu.dk/Projects/proj/nnctrl.html. These are freeware toolkits for system identification and control.

7.1 Artificial Neuron

The standard artificial neuron is a processing element whose output is calculated by multiplying its inputs by a weight vector, summing the results, and applying an activation function to the sum.

$$y = f\left[\sum_{k=1}^{n} x_k w_k + b_k\right]$$

The following figure depicts an artificial neuron with n inputs.

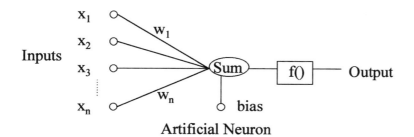

Artificial Neuron

The activation function could be one of many types. A linear activation function's output is simply equal to its input:

$$f(x) = x$$

```
x=[-5:0.1:5];
y=linear(x);
plot(x,y)
title('Linear Activation Function')
xlabel('x')
ylabel('Linear(x)')
```

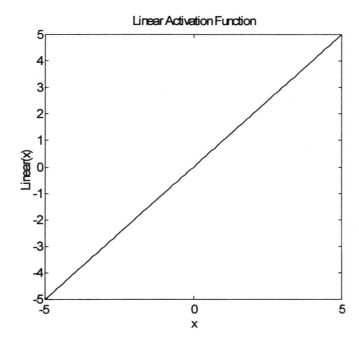

There are several types on non-linear activation functions. Differentiable, non-linear activation functions can be used in networks trained with backpropagation. The most common are the logistic function and the hyperbolic tangent function.

$$f(x) = \tanh(x) = \frac{e^x - e^{-x}}{e^x + e^{-x}}$$

Note that the output range of the logistic function is between -1 and 1.

```
x=[-3:0.1:3];
y=tanh(x);
plot(x,y)
title('Hyperbolic Tangent Activation Function')
xlabel('x')
ylabel('tanh(x)')
```

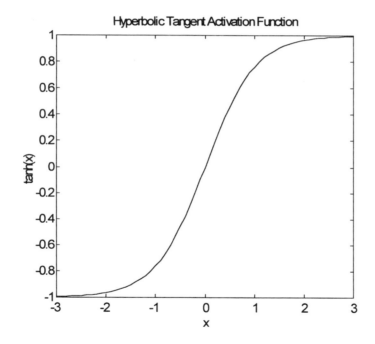

$$f(x) = logistic(x) = \frac{1}{1 + \exp(-\beta x)}$$

where β is the slope constant. We will always consider β to be one but it can be changed. Note that the output range of the logistic function is between 0 and 1.

```
x=[-5:0.1:5];
y=logistic(x);
plot(x,y)
title('Logistic Activation Function')
xlabel('x');ylabel('logistic(x)')
```

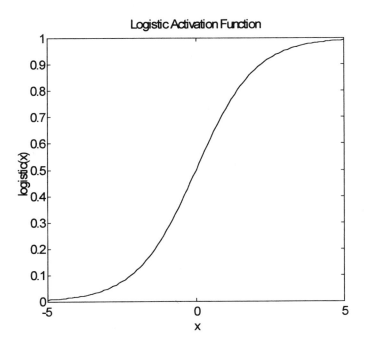

Non-differentiable non-linear activation functions are usually used as outputs of perceptrons and competitive networks. There are two type: the threshold function's output is either a 0 or 1 and the signum's output is a -1 or 1.

```
x=[-5:0.1:5];y=thresh(x);
plot(x,y);title('Thresh Activation Function')
xlabel('x');ylabel('thresh(x)')
```

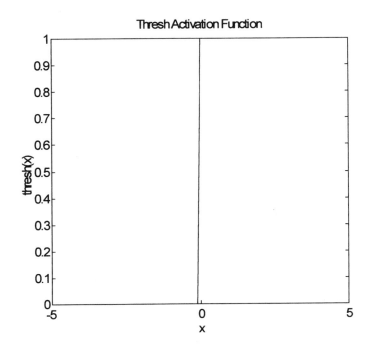

```
x=[-5:0.1:5];
y=signum(x);
plot(x,y)
title('Signum Activation Function')
xlabel('x')
ylabel('signum(x)')
```

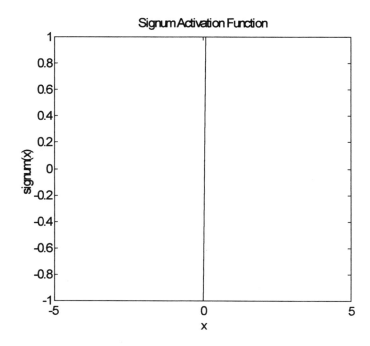

Note that the activation functions defined above can take a vector as input, and output a vector by performing the operation on each element of the input vector.

```
x=[-1 0 1];
linear(x)
logistic(x)
tanh(x)
thresh(x)
signum(x)

ans =
    -1     0     1
ans =
    0.2689    0.5000    0.7311
ans =
    -0.7616         0    0.7616
ans =
     0     1     1
ans =
    -1    -1     1
```

The output of a neuron is easily computed by using vector multiplication of the input and weights, and adding the bias. Suppose you have an input vector x=[2 4 6], and a weight matrix [.5 .25 .33] with a bias of -0.8. If the activation function is a hyperbolic tangent function, the output of the artificial neuron defined above is

```
x=[2 4 6]';
w=[0.5 -0.25 0.33];
b=-0.8;
y=tanh(w*x+b)
```

```
y =
    0.8275
```

7.2 Single Layer Neural Network

Neurons are grouped into layers and layers are grouped into networks to form highly interconnected processing structures. An input layer does no processing, it simply sends the inputs, modified by a weight, to each of the neurons in the next layer. This next layer can be a hidden layer or the output layer in a single layer design.

A bias is included in the neurons to allow the activation functions to be offset from zero. One method of implementing a bias is to use a dummy input node with a magnitude of 1. The weights connecting this dummy node to the next layer are the actual bias values.

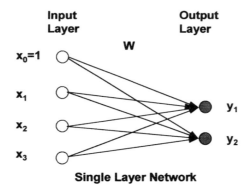
Single Layer Network

Suppose we have a single layer network with three input neurons and two output neurons as shown above. The outputs would be computed using matrix algebra in either of the two forms. The second form augments the input matrix with a dummy node and embeds the bias values into the weight matrix..

Form 1:

$$\mathbf{y} = \tanh(\mathbf{w} * \mathbf{x} + \mathbf{b}) = \tanh\left(\begin{bmatrix} 0.5 & -0.25 & 0.33 \\ 0.2 & -0.75 & -0.5 \end{bmatrix} \begin{bmatrix} 2 \\ 4 \\ 6 \end{bmatrix} + \begin{bmatrix} 0.4 \\ -1.2 \end{bmatrix}\right)$$

```
x=[2 4 6]';
w=[0.5 -0.25 0.33; 0.2 -0.75 -0.5];
```

```
b=[0.4 -1.2]';
y=tanh(w*x+b)

y =
    0.9830
   -1.0000
```

Form 2:

$$y = \tanh(\mathbf{w} * \mathbf{x}) = \tanh\left(\begin{bmatrix} 0.4 & 0.5 & -0.25 & 0.33 \\ -1.2 & 0.2 & -0.75 & -0.5 \end{bmatrix} \begin{bmatrix} 1 \\ 2 \\ 4 \\ 6 \end{bmatrix}\right)$$

```
x=[1 2 4 6]';
w=[0.4 0.5 -0.25 0.33; -1.2 0.2 -0.75 -0.5];
y=tanh(w*x)

y =
    0.9830
   -1.0000
```

7.3 Rosenblatt's Perceptron

The most simple single layer neuron is the perceptron and was developed by Frank Rosenblatt [1958]. A perceptron is a neural network composed of a single layer feed-forward network using threshold activation functions. Feed-forward means that all the interconnections between the layers propagate forward to the next layer. The figure below shows a single layer perceptron with two inputs and one output.

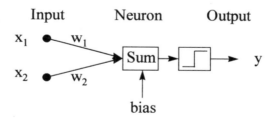

The simple perceptron uses the threshold activation function with a bias and thus has a binary output. The binary output perceptron has two possible outputs: 0 and 1. It is trained by supervised learning and can only classify input patterns that are linearly separable [Minsky 1969]. The next section gives an example of linearly separable data that the perceptron can properly classify.

Training is accomplished by initializing the weights and bias to small random values and then presenting input data to the network. The output (y) is compared to the target output (t=0 or t=1) and the weights are adapted according to Hebb's training rule [Hebb, 1949]:

"When the synaptic input and the neuron output are both active, the strength of the connection between the input and the output is enhanced."

This rule can be implemented as:

```
if       y = target    w = w;      % Correct output, no change.
  elseif y = 0         w = w+x;    % Target = 1, enhance strengths.
  else                 w = w-x;    % Target = 0, reduce strengths.
end
```

The bias is updated as a weight of a dummy node with an input of 1. The function trainpt1() implements this learning algorithm. It is called with:

`[w,b] = trainpt1(x,t,w,b);`

Assume the weight and bias values are randomly initialized and the following input and target output are given.

```
w = [.3  0.7];
b = [-0.8];
x = [1;-3];
t = [1];
```

the output is incorrect as shown:

```
y = thresh([w b]*[x ;1])

y =
    0
```

One learning cycle of the perceptron learning rule results in:

```
[w,b] = trainpt1(x,t,w,b)
y = thresh([w b]*[x ;1])

w =
    1.3000   -2.3000
b =
    0.2000
y =
    1
```

As can be seen, the weights are updated and the output now equals the target. Since the target was equal to 1, the weights corresponding to inputs with positive values were made stronger. For example, $x_1=1$ and w_1 changed from .3 to 1.3. Conversely, $x_2=-3$, and w_2 changed from 0.7 to -2.3; it was made more negative since the input was negative. Look at trainpt1 to see its implementation.

A single perceptron can be used to classify two inputs. For example, if $x_1 = [0,1]$ is to be classified as a 0 and $x_2 = [1\ -1]$ is to be classified as a 1, the initial weights and bias are chosen and the following training routine can be used.

```
x1=[0 1]';
x2=[1 -1]';
t=[0 1];
w=[-0.1 .8]; b=[-.5];
y1 = thresh([w b]*[x1 ;1])
y2 = thresh([w b]*[x2 ;1])


y1 =
    1
y2 =
    0
```

Neither output matches the target so we will train the network with first x_1 and then x_2.:

```
[w,b] = trainpt1(x1,t,w,b);
y1 = thresh([w b]*[x1 ;1])
y2 = thresh([w b]*[x2 ;1])
[w,b] = trainpt1(x2,t,w,b);
y1 = thresh([w b]*[x1 ;1])
y2 = thresh([w b]*[x2 ;1])


y1 =
    0
y2 =
    0
y1 =
    0
y2 =
    1
```

The network now correctly classifies the inputs. A better way of performing this training would be to modify trainpt1 so that it can take a matrix of input patterns such as $x = [x_1\ x_2]$. We will call this function trainpt(). Also, a function to simulate a perceptron with the inputs being a matrix of input patterns will be called percept().

```
w=[-0.1 .8]; b=[-.5];
y=percept(x,w,b)


y =
    0


[w,b] = trainpt(x,t,w,b)
y=percept(x,w,b)


w =
   -0.1000    0.8000
b =
```

```
            -0.5000
y =
             0
```

One training cycle results in the correct classification. This will not always be the case. It may take several training cycles, which are called epochs, to alter the weights enough to give the correct outputs. As long as the inputs are linearly separable, the perceptron will find a decision boundary which correctly divides the inputs. This proof is derived in many neural network texts and is called the perceptron convergence theorem [Hagan, Demuth and Beale, 1996]. The decision boundary is formed by the x,y pairs that solve the following equation:

w*x+b = 0

Let us now look at the decision boundaries before and after training for initial weights that correctly classify only one pattern.

```
x1=[0 0]';
x2=[1 -1]';
x=[x1 x2];
t=[0 1];
w=[-0.1 0.8]; b=[-0.5];
plot(x(1,:),x(2,:),'*')
axis([-1.5 1.5 -1.5 1.5]);hold on
X=[-1.5:.5:1.5];
Y=(-b-w(1)*X)./w(2);
plot(X,Y);hold;
title('Original Perceptron Decision Boundary')
```

```
Current plot released
```

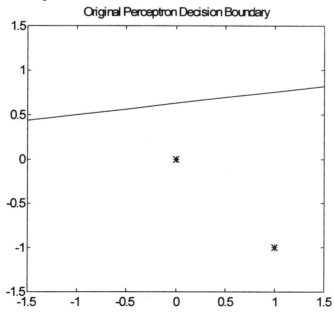

```
[w,b] = trainpt(x,t,w,b);
y=percept(x,w,b)
plot(x(1,:),x(2,:),'*')
axis([-1.5 1.5 -1.5 1.5]);
hold on
X=[-1.5:.5:1.5]; Y=(-b-w(1)*X)./w(2);
plot(X,Y);
hold
title('Perceptron Decision Boundary After One Epoch')

y =
     1     1
Current plot released
```

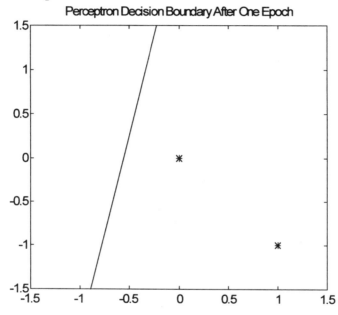

Note that after one epoch, still only one pattern is correctly classified.

```
[w,b] = trainpt(x,t,w,b);
y=percept(x,w,b)
plot(x(1,:),x(2,:),'*')
axis([-1.5 1.5 -1.5 1.5])
hold on
X=[-1.5:.5:1.5]; Y=(-b-w(1)*X)./w(2);
plot(X,Y)
hold
title('Perceptron Decision Boundary After Two Epochs')

y =
     0     1
Current plot released
```

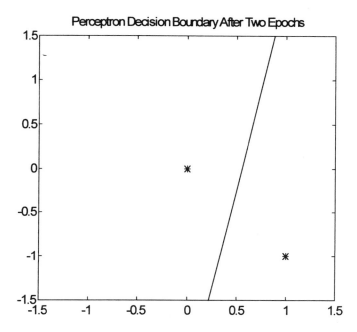

Note that after two epochs, both patterns are correctly classified.

The perceptron can also be used to classify several linearly separable patterns. The function percept() will now be modified to train until the patterns are correctly classified or until 20 epochs.

```
x=[0 -.3 .5 1;-.4 -.2 1.3 -1.3];
t=[0 0 1 1]
w=[-0.1 0.8]; b=[-0.5];
y=percept(x,w,b)
plot(x(1,1:2),x(2,1:2),'*')
hold on
plot(x(1,3:4),x(2,3:4),'+')
axis([-1.5 1.5 -1.5 1.5])
X=[-1.5:.5:1.5]; Y=(-b-w(1)*X)./w(2);
plot(X,Y)
hold
title('Original Perceptron Decision Boundary')

t =
     0     0     1     1
y =
     0     0     1     0
Current plot released
```

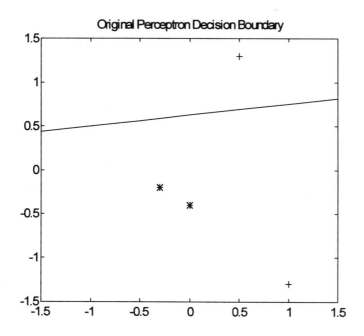

The original weight and bias values misclassifies pattern number 4.

```
[w,b] = trainpti(x,t,w,b)
t
y=percept(x,w,b)
plot(x(1,1:2),x(2,1:2),'*')
hold on
plot(x(1,3:4),x(2,3:4),'+')
axis([-1.5 1.5 -1.5 1.5])
X=[-1.5:.5:1.5]; Y=(-b-w(1)*X)./w(2);
plot(X,Y)
hold
title('Final Perceptron Decision Boundary')

Solution found  in 5 epochs.
w =
    2.7000    0.5000
b =
   -0.5000
t =
       0       0       1       1
y =
       0       0       1       1
Current plot released
```

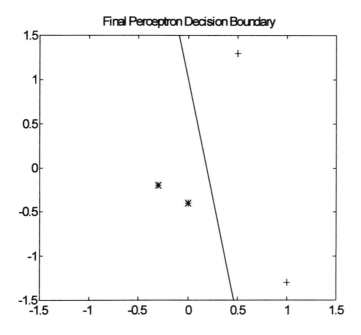

After 5 epochs, all 5 inputs are correctly classified.

7.4 Separation of Linearly Separable Variables

A two input perceptron can separate a plane into two sections because its transfer equation can be rearranged to form the equation for a line. In a three dimensional problem, the equation would define a plane and in higher dimensions it would define a hyperplane.

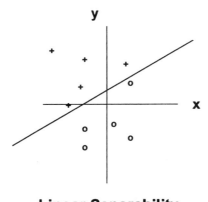

Linear Separability

Note that the decision boundary is always orthogonal to the weight matrix. Suppose we have a two input perceptron with weights = [1 2] and a bias equal to 1. The decision boundary is defined as:

$$y = -\frac{w_x}{w_y}x - \frac{b}{w_y} = -\frac{1}{2}x - \frac{1}{2} = -0.5x - 0.5$$

which is orthogonal to the weight vector [1 2]. In the figure below, the more vertical line is the decision boundary and the more horizontal line is the weight vector extended to meet the decision boundary.

```
w=[1 2]; b=[1];
x=[-2.5:.5:2.5]; y=(-b-w(1)*x)./w(2);
plot(x,y)
text(.5,-.65,'Decision Boundary');
grid
title('Perceptron Decision Boundary')
xlabel('x');ylabel('y');
hold on
plot([w(1) -w(1)],[w(2) -w(2)])
text(.5,.7,'Weight Vector');
axis([-2 2 -2 2]);
hold
```

```
Current plot released
```

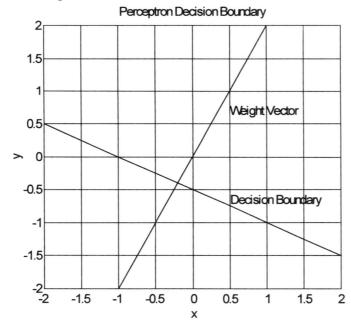

If the inputs need to be classified into three or four classes, a two neuron perceptron can be used. The outputs can be coded to one of the pattern classifications, and two lines can separate the classification regions. In the following example, each of the inputs will be classified into one of the three binary classes: [0 1], [1 0], and [0 0]. . The weights can be defined as a matrix, the bias is a vector, and two lines are formed.

```
x=[0 -.3 .5 1;-.4 -.2 1.3 -1.3];          % Input vectors
```

```
t=[0 0 1 1; 0 1 0 0]                    % Target vectors
w=[-0.1 0.8; 0.2 -0.9]; b=[-0.5;0.3];    % Weights and biases
y=percept(x,w,b)                         % Initial classifications

t =
     0     0     1     1
     0     1     0     0
y =
     0     0     1     0
     1     1     0     1
```

Two of the patterns (t_1 and t_4) are incorrectly classified.

```
[w,b] = trainpti(x,t,w,b)
t
y=percept(x,w,b)

Solution found  in 6 epochs.
w =
    2.7000    0.5000
   -2.2000   -0.7000
b =
   -0.5000
   -0.7000
t =
     0     0     1     1
     0     1     0     0
y =
     0     0     1     1
     0     1     0     0
```

The perceptron learning algorithm was able to define lines that separated the input patterns into their target classifications. This is shown in the following figure.

```
plot(x(1,1),x(2,1),'*')
hold on
plot(x(1,2),x(2,2),'+')
plot(x(1,3:4),x(2,3:4),'o')
axis([-1.5 1.5 -1.5 1.5])
X1=[-1.5:.5:1.5]; Y1=(-b(1)-w(1,1)*X1)./w(1,2);
plot(X1,Y1)
X2=[-1.5:.5:1.5]; Y2=(-b(2)-w(2,1)*X2)./w(2,2);
plot(X2,Y2)
hold
title('Perceptron Decision Boundaries')
text(-1,.5,'A'); text(-.3,.5,'B'); text(.5,.5,'C');

Current plot released
```

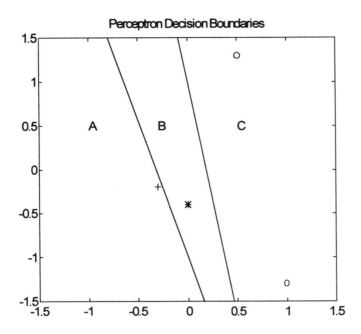

The simple single layer perceptron can separate linearly separable inputs but will fail if the inputs are not linearly separable. One such example of linearly non-separable inputs is the exclusive-or (XOR) problem. Linearly non-separable patterns, such as those of the XOR problem, can be separated with multilayer networks. A two input, one hidden layer of two neurons, one output network defines two lines in the two dimensional space. These two lines can classify points into groups that are not linearly separable with one line.

Perceptrons are limited in that they can only separate linearly separable patterns and that they have a binary output. Many of the limitations of the simple perceptron can be solved with multi-layer architectures, non-binary activation functions, and more complex training algorithms. Multilayer perceptrons with threshold activation functions are not that useful because they can't be trained with the perceptron learning rule and since the functions are not differentiable, they can't be trained with gradient descent algorithms. Although if the first layer is randomly initialized, the second layer may be trained to classify linearly non-separable classes (MATLAB Neural Networks Toolbox).

The **Adaline** (adaptive linear) network, also called a **Widrow Hoff** network, developed by Bernard Widrow and Marcian Hoff [1960], is composed of one layer of linear transfer functions, as opposed to threshold transfer functions, and thus has a continuous valued output. It is trained with supervised training by the Delta Rule which will be discussed in Chapter 8.

7.5 Multilayer Neural Network

Neural networks with one or more hidden layers are called multilayer neural networks or multilayer perceptrons (MLP). Normally, each hidden layer of a network uses the same type of activation function. The output activation function is either sigmoidal or linear.

The output of a sigmoidal neuron is constrained [-1 1] for a hyperbolic tangent neuron and [0 1] for a logarithmic sigmoidal neuron. A linear output neuron is not constrained and can output a value of any magnitude.

It has been proven that the standard feedforward multilayer perceptron (MLP) with a single non-linear hidden layer (sigmoidal neurons) can approximate any continuous function to any desired degree of accuracy over a compact set [Cybenko 1989, Hornick 1989, Funahashi 1989, and others], thus the MLP has been termed a universal approximator. Haykin [1994] gives a very concise overview of the research leading to this conclusion.

What this proof does not say is how many hidden layer neurons would be needed, and if the weight matrix that corresponds to that error goal can be found. It may be computationally limiting to train such a network since the size of the network is dependent on the complexity of the function and the range of interest.

For example, a simple non-linear function:

$$f(\mathbf{x}) = x_1 \cdot x_2$$

requires many nodes if the ranges of x_1 and x_2 are very large.

In order to be a universal approximation, the hidden layer of a multilayer perceptron is usually a sigmoidal neuron. A linear hidden layer is rarely used because any two linear transformations

$$\mathbf{h} = \mathbf{W1}\, \mathbf{x}$$
$$\mathbf{y} = \mathbf{W2}\, \mathbf{h}$$

where **W1** and **W2** are transformation matrices that transform the mx1 vector **x** to **h** and **h** to **y**, can be represented as one linear transformation

$$\mathbf{W} = \mathbf{W1}\, \mathbf{W2}$$
$$\mathbf{y} = \mathbf{W1}\ \mathbf{W2}\, \mathbf{x} = \mathbf{W}\, \mathbf{x}$$

where **W** is a matrix that performs the transformation from **x** to **y**. The following figure shows the general multilayer neural network architecture.

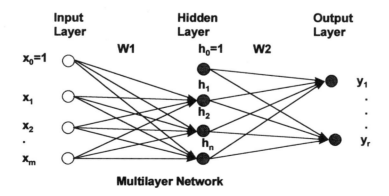

Multilayer Network

The output for a single hidden layer MLP with three inputs, three hidden hyperbolic tangent neurons and two linear output neurons can be calculated using matrix algebra.

$$y = w_2 * \tanh(w_1 * x + b_1) + b_2$$

$$= \begin{bmatrix} 0.5 & -0.25 & 0.33 \\ 0.2 & -0.75 & -0.5 \end{bmatrix} \tanh\left(\begin{bmatrix} 0.2 & -0.7 & 0.9 \\ 2.3 & 1.4 & -2.1 \\ 10.2 & -10.2 & 0.3 \end{bmatrix} \begin{bmatrix} 2 \\ 4 \\ 6 \end{bmatrix} + \begin{bmatrix} 0.5 \\ 0.2 \\ -0.8 \end{bmatrix}\right) + \begin{bmatrix} 0.4 \\ -1.2 \end{bmatrix}$$

By using dummy nodes and embedding the biases into the weight matrix we can use a more compact notation:

$$y = w_2 * [1 \,; \tanh(w_1 * [1\,;x])]$$

$$= \begin{bmatrix} 0.4 & 0.5 & -0.25 & 0.33 \\ -1.2 & 0.2 & -0.75 & -0.5 \end{bmatrix} \left\{ \begin{matrix} 1 \\ \tanh\left(\begin{bmatrix} 0.5 & 0.2 & -0.7 & 0.9 \\ 0.2 & 2.3 & 1.4 & -2.1 \\ -0.8 & 10.2 & -10.2 & 0.3 \end{bmatrix} \begin{bmatrix} 1 \\ 2 \\ 4 \\ 6 \end{bmatrix}\right) \end{matrix} \right\}$$

```
x=[2 4 6]';
w1=[0.2 -0.7 0.9; 2.3 1.4 -2.1; 10.2 -10.2 0.3];
w2=[0.5 -0.25 0.33; 0.2 -0.75 -0.5];
b1=[0.5 0.2 -0.8]';
b2=[0.4 -1.2]';
y=w2*tanh(w1*x+b1)+b2

y =
    0.8130
    0.2314
```

or

```
x=[2 4 6]';
w1=[0.5 0.2 -0.7 0.9; 0.2 2.3 1.4 -2.1; -0.8 10.2 -10.2 0.3];
w2=[0.4 0.5 -0.25 0.33; -1.2 0.2 -0.75 -0.5];
y=w2*[1;tanh(w1*[1;x])]

y =
    0.8130
    0.2314
```

Chapter 8 Backpropagation and Related Training Paradigms

Backpropagation (BP) is a general method for iteratively solving for a multilayer perceptrons' weights and biases. It uses a steepest descent technique which is very stable when a small learning rate is used, but has slow convergence properties. Several methods for speeding up BP have been used including momentum and a variable learning rate.

Other methods of solving for the weights and biases involve more complex algorithms. Many of these techniques are based on Newton's method, but practical implementations usually use a combination of Newton's method and steepest descent. The most popular second order method is the Levenberg [1944] and Marquardt [1963] technique. The scaled conjugate gradient technique [Moller 1993] is also very popular because it less memory intensive than Levenberg Marquardt and more powerful than gradient descent techniques. These methods will not be implemented in this supplement although Levenberg Marquardt is implemented in The MathWork's Neural Network Toolbox..

8.1 Derivative of the Activation Functions

The chain rule that is used in deriving the BP algorithm necessitates the computation of the derivative of the activation functions. For logistic, hyperbolic tangent, and linear functions; the derivatives are as follows:

Linear $\quad \Phi(I) = I \quad\quad\quad \dot{\Phi}(I) = 1$

Logistic $\quad \Phi(I) = \dfrac{1}{1+e^{-\alpha I}} \quad\quad \dot{\Phi}(I) = \alpha \Phi(I)(1-\Phi(I))$

Tanh $\quad \Phi(I) = \dfrac{e^{\alpha I} - e^{-\alpha I}}{e^{\alpha I} + e^{-\alpha I}} \quad\quad \dot{\Phi}(I) = \alpha(1-\Phi(I)^2)$

Alpha is called the slope parameter. Usually alpha is chosen to be 1 but other slopes may be used. This formulation for the derivative makes the computation of the gradient more efficient since the output $\Phi(I)$ has already been calculated in the forward pass. A plot of the logistic function and its derivative follows.

```
x=[-5:.1:5];
y=1./(1+exp(-x));
dy=y.*(1-y);
subplot(2,1,1)
plot(x,y)
title('Logistic Function')
subplot(2,1,2)
plot(x,dy)
title('Derivative of Logistic Function')
```

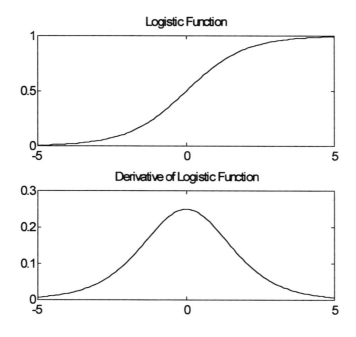

As can be seen, the highest gradient is at I=0. Since the speed of learning is partially dependent on the size of the gradient, the internal activation of all neurons should be kept small to expedite training. This is why we scale the inputs and initialize weights to small random values.

Since there is no logistic function in MATLAB, we will write a m-file function to perform the computation.

```
function [y] = logistic(x);
% [y] = logistic(x)
% Returns the result of applying the logistic operator to the input x.
% x  : the input
% y  : the result
y=1./(1+exp(-x));
```

8.2 Backpropagation for a Multilayer Neural Network

The backpropagation algorithm is an optimization technique designed to minimize an objective function [Werbos 1974]. The most commonly used objective function is the squared error which is defined as:

$$\varepsilon^2 = \left[T_q - \Phi_{qk} \right]^2.$$

The network syntax is defined as in the following figure:

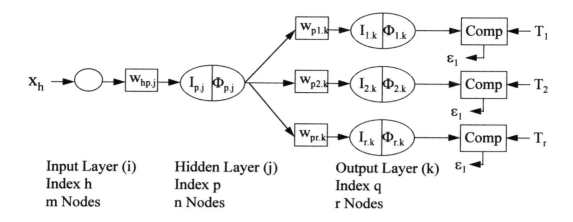

Input Layer (i)   Hidden Layer (j)   Output Layer (k)
Index h   Index p   Index q
m Nodes   n Nodes   r Nodes

In this notation, the layers are labeled i, j, and k; with m, n, and r neurons respectively; and the neurons in each layer are indexed h, p, and q respectively.

   x = input value
   T = target output value
   w = weight value
   I = internal activation
   Φ = neuron output
   ε = error term

The outputs for a two layer network with both layers using a logistic activation function are calculated by the equation:

$$\Phi = \text{logistic}\{\mathbf{w2} * [\text{logistic}(\mathbf{w1} * \mathbf{x} + \mathbf{b1})] + \mathbf{b2}\}$$

where: **w1** = first layer weight matrix
             w2 = second layer weight matrix
             b1 = first layer bias vector
             b2 = second layer bias vector

The input vector can be augmented with a dummy node representing the bias input. This dummy input of 1 is multiplied by a weight corresponding to the bias value. This results in a more compact representation of the above equation:

$$\Phi = \text{logistic}\left\{\mathbf{W2} * \begin{bmatrix} 1 \\ \text{logistic}(\mathbf{W1} * \mathbf{X}) \end{bmatrix}\right\}$$

where   **X** = [1 **x**]'   % Augmented input vector.
          W1 = [**b1 w1**]
          W2 = [**b2 w2**]

Note that a dummy hidden node (=1) also needs to be inserted into the equation.

For a two input network with two hidden nodes and one output node we have:

$$\mathbf{x} = \begin{bmatrix} 1 \\ x_1 \\ x_2 \end{bmatrix} \qquad \mathbf{W1} = \begin{bmatrix} b_{11} & w_{11} & w_{21} \\ b_{12} & w_{12} & w_{22} \end{bmatrix} \qquad \mathbf{W2} = \begin{bmatrix} b_2 & w_{11} & w_{12} \end{bmatrix}$$

As an example, consider a network that has 2 inputs, one hidden layer of 2 logistic neurons, and 1 logistic output neuron (same as in the text). After we define the initial weight and bias matrices we combine them to be used in the MATLAB code and calculate the output value for the initial weights.

```
x = [0.4;0.7];              % rows = # of inputs = 2
                            % columns = # of patterns = 1
w1 = [0.1 -0.2; 0.4 0.2];   % rows = # of hidden nodes = 2
                            % columns = # of inputs = 2
w2 = [0.2 -0.5];            % rows = # of outputs = 1
                            % columns = # of hidden nodes = 2
b1=[-0.5;-0.2];             % rows = number of hidden nodes = 2
b2=[-0.6];                  % rows = number of output nodes = 1

X = [1;x]                   % Augmented input vector.
W1 = [b1 w1]
W2 = [b2 w2]
output=logistic(W2*[1;logistic(W1*X)])

X =
    1.0000
    0.4000
    0.7000
W1 =
   -0.5000    0.1000   -0.2000
   -0.2000    0.4000    0.2000
W2 =
   -0.6000    0.2000   -0.5000
output =
    0.3118
```

8.2.1 Weight Updates

The output layer weights are changed in proportion to the negative gradient of the squared error with respect to the weights. These weight changes can be calculated using the chain rule. The following is the derivation for a two layer network with each layer having logistic activation functions. Note that the target outputs can only have a range of [0 1] for a network with a logistic output layer.

$$\Delta w_{pq.k} = -\eta_{p.q} \frac{\partial \varepsilon^2}{\partial w_{pq.k}}$$

$$= -\eta_{p.q} \cdot \frac{\partial \varepsilon^2}{\partial \Phi_{q.k}} \cdot \frac{\partial \Phi_{q.k}}{\partial I_{q.k}} \cdot \frac{\partial I_{q.k}}{\partial w_{pq.k}}$$

$$= -\eta_{p.q} \cdot (-2)[T_q - \Phi_{q.k}] \cdot \Phi_{q.k}[1 - \Phi_{q.k}] \cdot \Phi_{p.j}$$

$$= -\eta_{p.q} \cdot \delta_{pq.k} \cdot \Phi_{p.j}$$

and

$$\delta_{pq.k} = 2[T_q - \Phi_{q.k}]\Phi_{q.k}[1 - \Phi_{q.k}]$$

The weight update equation for the output neurons is:

$$w_{pq.k}(N+1) = w_{pq.k}(N) - \eta_{p.q} \cdot \delta_{pq.k} \cdot \Phi_{p.j}$$

The output of the hidden layer of the above network is Φ_{pj}=h=[0.48 0.57]. The target output is T = [0.7]. The update for the output layer weights is calculated by first propagating forward through the network to calculate the error terms:

```
h = logistic(W1*X);          % Hidden node output.
H = [1;h];                   % Augmented hidden node output.
t = [0.1];                   % Target output.
Out_err = t-logistic(W2*[1;logistic(W1*X)])

Out_err =
   -0.2118
```

Next, the gradient vector for the output weight matrix is calculated.

```
output=logistic(W2*[1;logistic(W1*X)])
delta2=output.*(1-output).*Out_err    % Derivative of logistic function.

output =
    0.3118
delta2 =
   -0.0455
```

And lastly, the weights are updated with the learning rate α =0.5.

```
lr = 0.5;
del_W2 = 2*lr*H'*delta2      % Change in weight.
new_W2 = W2+del_W2           % Weight update.

del_W2 =
   -0.0455   -0.0161   -0.0239
new_W2 =
```

```
      -0.6455    0.1839    -0.5239

Out_err = t-logistic(new_W2*[1;logistic(W1*X)])    % New output error)

Out_err =
    -0.1983
```

We see that by updating the output weight matrix for one iteration of training, the error is reduced from 0.2118 to 0.1983.

8.2.2 Hidden Layer Weight Updates

The hidden layer outputs have no target values. Therefore, a procedure is used to backpropagate the output layer errors to the hidden layer neurons in order to modify their weights to minimize the error. To accomplish this, we start with the equation for the gradient with respect to the weights and use the chain rule.

$$\Delta w_{hp.j} = -\eta_{h.p} \frac{\partial \varepsilon^2}{\partial w_{hp.j}}$$

$$= -\eta_{h.p} \sum_{q=1}^{r} \frac{\partial \varepsilon^2}{\partial w_{hp.j}}$$

$$= -\eta_{h.p} \sum_{q=1}^{r} \frac{\partial \varepsilon_q^2}{\partial \Phi_{q.k}} \cdot \frac{\partial \Phi_{q.k}}{\partial I_{q.k}} \cdot \frac{\partial I_{q.k}}{\partial \Phi_{p.j}} \cdot \frac{\partial \Phi_{p.j}}{\partial I_{p.j}} \cdot \frac{\partial I_{p.j}}{\partial w_{hp.j}}$$

$$\frac{\partial \varepsilon_q^2}{\partial \Phi_{q.k}} = (-2)\left[T_q - \Phi_{q.k}\right]$$

$$\frac{\partial \Phi_{q.k}}{\partial I_{q.k}} = \alpha \Phi_{q.k}\left[1 - \Phi_{q.k}\right]$$

$$\frac{\partial I_{q.k}}{\partial \Phi_{p.j}} = w_{pq.k}$$

$$\frac{\partial \Phi_{p.j}}{\partial I_{p.j}} = \alpha \Phi_{p.j}\left[1 - \Phi_{p.j}\right]$$

$$\frac{\partial I_{p.j}}{\partial w_{hp.j}} = x_h$$

resulting in:

$$\frac{\partial \varepsilon^2}{\partial w_{hp.j}} = \sum_{q=1}^{r}(-2)\left[T_q - \Phi_{q.k}\right] \cdot \alpha\, \Phi_{q.k}\left[1-\Phi_{q.k}\right] \cdot w_{pq.k} \cdot \alpha\, \Phi_{p.j}\left[1-\Phi_{p.j}\right]x_h$$

$$= \sum_{q=1}^{r} \delta_{pq.k} \cdot w_{pq.k} \cdot \alpha\, \Phi_{p.j}\left[1-\Phi_{p.j}\right]x_h$$

$$\delta_{hp.j} = \delta_{pq.k} w_{pq.k} \frac{\partial \Phi_{p.j}}{\partial I_{p.j}}$$

$$w_{hp.j}(N+1) = w_{hp.j}(N) - \eta_{hp} x_h \delta_{hp.j}$$

First we must backpropagate the gradient terms back to the hidden layer:

```
[numout,numhid]=size(W2);
delta1=delta2.*h.*(1-h).*W2(:,2:numhid)'
```

```
delta1 =
   -0.0021
    0.0057
```

Now we calculate the hidden layer weight change. Note that we don't propagate back through the dummy node.

```
del_W1 = 2*lr*delta1*X'
new_W1 = W1+del_W1

del_W1 =
   -0.0021   -0.0008   -0.0015
    0.0057    0.0023    0.0040
new_W1 =
   -0.5021    0.0992   -0.2015
   -0.1943    0.4023    0.2040
```

Now we calculate the new output value.

```
output = logistic(new_W2*[1;logistic(new_W1*X)])

output =
    0.2980
```

The new output is 0.298 which is closer to the target of 0.1 than was the original output of 0.3118. The magnitude of the learning rate affects the convergence and stability of the training. This will be discussed in greater detail in the adaptive learning rate section.

8.2.3 Batch Training

The type of training used in the previous section is called sequential training. During sequential training, the weights are updated after each presentation of a pattern. This method may be more stochastic when the patterns are chosen randomly and may reduce the chance of getting stuck in a local minima. During batch training all of the training patterns are processed before a weight update is made. Suppose the training set consists of four patterns (z=4). Now we have four output patterns from the hidden layer of the network and four target outputs.

```
x = [0.4 0.8 1.3 -1.3;0.7 0.9 1.8 -0.9];
t = [0.1 0.3 0.6 0.2];
[inputs,patterns] = size(x);
[outputs,patterns] = size(t);
W1 = [0.1 -0.2 0.1; 0.4 0.2 0.9];    % rows = # of hidden nodes = 2
                                      % columns = # of inputs +1 = 3
W2 = [0.2 -0.5 0.1];                  % rows = # of outputs = 1
                                      % columns = # of hidden nodes+1 = 3
X = [ones(1,patterns); x];            % Augment with bias dummy node.
h = logistic(W1*X);
H = [ones(1,patterns);h];
e = t-logistic(W2*H)


e =
   -0.4035   -0.2065    0.0904   -0.2876
```

The sum of squared error is:

```
SSE = sum(sum(e.^2))

SSE =
    0.2963
```

Next, the gradient vector is calculated:

```
output = logistic(W2*H)
delta2 = output.*(1-output).*e

output =
    0.5035    0.5065    0.5096    0.4876
delta2 =
   -0.1009   -0.0516    0.0226   -0.0719
```

And lastly, the weights are updated with a learning rate α =0.5:

```
lr = 0.5;
del_W2 = 2*lr* delta2*H'
new_W2 = W2+ del_W2

del_W2 =
   -0.2017   -0.1082   -0.1208
new_W2 =
   -0.0017   -0.6082   -0.0208
```

The new sum of squared error is calculated as:

```
e = t- logistic(new_W2*H);
```

```
SSE = sum(sum(e.^2))

SSE =
    0.1926
```

The SSE has been reduced from 0.2963 to 0.1926 by just changing the output layer weights and biases.

To change the hidden layer weight matrix we must backpropagate the gradient terms back to the hidden layer. Note that we can't backpropagate through a dummy node, so only the weight portion of W2 is used.

```
[numout,numhidb] = size(W2);
delta1 = h.*(1-h).*(W2(:,2:numhidb)'*delta2)

delta1 =
    0.0126    0.0065   -0.0028    0.0088
   -0.0019   -0.0008    0.0002   -0.0016
```

Now we calculate the hidden layer weight change.

```
del_W1 = 2*lr*delta1*X'
new_W1 = W1+del_W1

del_W1 =
    0.0250   -0.0049    0.0016
   -0.0041    0.0009   -0.0003
new_W1 =
    0.1250   -0.2049    0.1016
    0.3959    0.2009    0.8997

h = logistic(new_W1*X);
H = [ones(1,patterns);h];
e = t-logistic(new_W2*H);
SSE = sum(sum(e.^2))

SSE =
    0.1917
```

The new SSE is 0.1917 which is less than the SSE of 0.1926 so the change in hidden layer weights reduced the SSE slightly more.

8.2.4 Adaptive Learning Rate

In the previous examples a fixed learning rate was used. When training a neural network iteratively, it is more efficient to use an adaptive learning rate. The learning rate can be thought of as the size of a step down the error gradient. If very small steps are taken, you are guaranteed to fine an error minimum, but this may take a very long time. Larger steps may result in unstable learning since you may step over a minima. To speed training and still have stability, a heuristic method is used to determine the step size.

The heuristic rule states:

If training is "went well" (error decreased) then increase the step size.
   lr=lr*1.1
If training is "poor" (error increased) then decrease the step size.
   lr=lr*0.5
Only update the weights if the error decreased.

Using an adaptive learning rate allows the training procedure to quickly move across large error plateaus and slowly descend through tortuous error paths. This results in training that is somewhat optimized to increase learning speed while remaining stable.

8.2.5 The Backpropagation Training Cycle

The training procedure discussed above is applied iteratively for a certain number of cycles or until a specified error goal is met. Before starting training, weights are initialized to small random values and inputs are scaled to small values of similar magnitude to reduce the chance of prematurely saturating the sigmoidal neurons and thus slowing training. Input scaling and weight initialization will be covered in subsequent sections.

```
x = [0.4 0.8 1.3 -1.3;0.7 0.9 1.8 -0.9];
t = [0.1 0.3 0.6 0.2];
[inputs,patterns] = size(x);
[outputs,patterns] = size(t);
hidden=6;
W1=0.1*ones(hidden,inputs+1);   % Initialize to matrix of 0.1's.
W2=0.1*ones(outputs,hidden+1);  % Initialize to matrix of 0.1's.
maxcycles=200;SSE_Goal=0.1;
lr=0.5;SSE=zeros(1,maxcycles);
X=[ones(1,patterns); x];   % Augment inputs with bias dummy node.
for i=1:maxcycles
   h = logistic(W1*X);
   H=[ones(1,patterns);h];
   e=t-logistic(W2*H);
   SSE(i)= sum(sum(e.^2));
   if SSE(i)<SSE_Goal; break;end
   output = logistic(W2*H);
   delta2= output.*(1-output).*e;
   del_W2= 2*lr* delta2*H';
   W2 = W2+ del_W2;
   delta1 = h.*(1-h).*(W2(:,2:hidden+1)'*delta2);
   del_W1 = 2*lr*delta1*X';
   W1 = W1+del_W1;
end;clf
semilogy(nonzeros(SSE));
title('Backpropagation Training');
xlabel('Cycles');ylabel('Sum of Squared Error')
if i<200;fprintf('Error goal reached in %i cycles.',i);end

Error goal reached in 116 cycles.
```

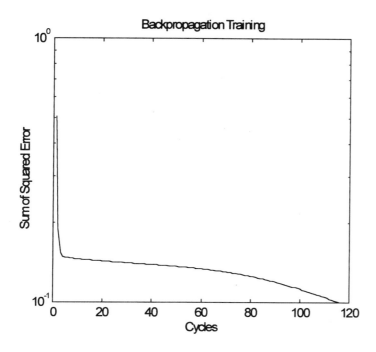

You can try different numbers of hidden nodes in the above example. The relationship between the number of hidden nodes and the cycles to reach the error goal is shown below. Note that the weights are usually initialized to random numbers as discussed in section 8.3. When the weights are randomly chosen, the number of training cycles varies. This is due to the different initial position on the error surface. In fact, sometimes a network will not train to a desired error goal with one set of initial weights, due to getting trapped in a local minima, but will train with a different set of initial weights.

Number of Hidden Nodes	Training Cycles to Error Goal
1	106
2	95
3	97
4	102
5	109
6	116

Each training set will have its own relationship between training cycles and the number of hidden nodes. The results shown above are fairly typical. After a point where enough free parameters are given to the network to model the function, the addition of hidden nodes complicates the error surface and the number of training cycles may increase.

8.3 Scaling Input Vectors

Training data is scaled for two major reasons. First, input data is usually scaled to give each input equal importance and to prevent premature saturation of sigmoidal activation functions. Secondly, output or target data is scaled if the output activation functions have a limited range and the unscaled targets do not match that range.

There are two popular types of input scaling: linear scaling and z-score scaling. Linearly scaling transforms the data into a new range which is usually 0.1 to 0.9. If the training patterns are in a form such that the columns are inputs and the rows are patterns, a MATLAB function to perform linear scaling is:

```
function [y,slope,int]=scale(x,slope,int)
%  [x,m,b]=scale(x,m,b)
%
%  Linear scale the data between .1 and .9
%
%      y = m*x + b
%
%      x = data
%      m = slope
%      b = y intercept
%
[nrows,ncols]=size(x);

if nargin == 1
   del = max(x)-min(x);  % calculate slope and intercept
   slope = .8./del;
   int = .1 - slope.*min(x);
end

y = (ones(nrows,1)*slope).*x + ones(nrows,1)*int;
```

The function returns the scaled inputs y and the scale parameters: slope and int. Each column of inputs has a maximum and minimum value of 0.9 and 0.1 respectively. The scaling parameters are returned so that other input data may be transformed using the same terms. A network that is trained with scaled inputs must always have its inputs scaled using the same scaling parameters.

This scaling function can also be used for scaling target values. Outputs are usually scaled to 0.9 to 0.1 when logistic output activation functions are used. This keeps the training algorithm from trying to force outputs beyond the range of the function.

Another method of scaling, called z-score or mean center unit variance scaling is also frequently used. This method subtracts the mean of each input from each column and then divides by the variance. This centers all the patterns of each data type around 0 and gives them a unit variance. The MATLAB function to perform z-score scaling is:

```
function [y,meanval,stdval] = zscore(x, meanaval,stdval)
%
%  [y,mean,std] = zscore(x, mean_in,std_in)
```

```
%
% Mean center the data and scale to unit variance.
% If number of inputs is one, calculate the mean and standard deviation.
% If the number if inputs is three, use the calculated mean and SD.
%
[nrows,ncols]=size(x);

if nargin == 1
   meanval = mean(x);   % calculate mean values
end

y = x - ones(nrows,1)*meanval;  % subtract off mean

if nargin == 1
   stdval = std(y); % calculate the SD
end

y = y ./ (ones(nrows,1)*stdval); % normalize to unit variance
```

An example of the z-score scaling function is:

```
x=[1 2;30 21;-1 -10;8 34]
[y,slope, int]=scale(x)
[y,meanval,stdval]=zscore(x)

x =
     1     2
    30    21
    -1   -10
     8    34
y =
    0.1516    0.3182
    0.9000    0.6636
    0.1000    0.1000
    0.3323    0.9000
slope =
    0.0258    0.0182
int =
    0.1258    0.2818
y =
   -0.5986   -0.4983
    1.4436    0.4727
   -0.7394   -1.1115
   -0.1056    1.1370
meanval =
    9.5000   11.7500
stdval =
   14.2009   19.5683
```

If a network is trained with scaled data and new data is presented to the network, it must first be scaled using the same scaling factors. In this case, the scaling functions are called with three variables and only the scaled data is passed back.

```
x_new=[-2 4;-.3 12; 9 -10]
[y]=scale(x_new, slope, int)
[y]=zscore(x_new,meanval,stdval)
```

```
x_new =
   -2.0000    4.0000
   -0.3000   12.0000
    9.0000  -10.0000
y =
    0.0742    0.3545
    0.1181    0.5000
    0.3581    0.1000
y =
   -0.8098   -0.3960
   -0.6901    0.0128
   -0.0352   -1.1115
```

8.4 Initializing Weights

As mentioned above, the initial weights should be selected to be small random values in order to prevent premature saturation of the sigmoidal activation functions. The most common method is to use the random number generator and pass it the number of inputs plus 1 and the number of hidden nodes for the first hidden layer weight matrix W1 and pass it the number of outputs and hidden nodes plus 1 for the output weight matrix W2. One is added to the number of inputs in W1 and to hidden in W2 to account for the bias. To make the weights somewhat smaller, the resulting random weight matrix is multiplied by 0.5.

```
W1=0.5*randn(2,3)
```

```
W1 =
    0.5825    0.0375   -0.3483
    0.3134    0.1758    0.8481
```

We are trying to limit the internal activation of the neurons during training to a high gradient region. This region is between -2.5 and 2.5 for a hyperbolic tangent neuron and -5 to +5 for a logistic function.

```
plot([-8:.1:8], logistic([-8:.1:8]))
title('Logistic Activation Function')
```

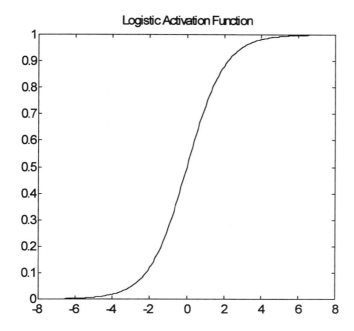

The sum of the inputs times their weights should be in this high gradient region for efficient training. Scaling the inputs to small values helps, but the weights should also be made small, random, and centered around 0.

8.5 Creating a MATLAB Function for Backpropagation

In this section, a MATLAB script will be discussed that performs the necessary operations to define and train a multilayer perceptron with a backpropagation algorithm. Up to this point, the two layer networks used logistic activation functions in each layer. This limits the network output to the interval [0 1]. Since most problems have targets outside of this range, the training function backprop() will use a linear output layer that allows targets of any magnitude to be reached.

The backpropagation MATLAB script that sets up the network architecture and training parameters is called bptrain. To use this script you must first have training data (x,t) saved in a file (data.mat). The following defines the format for these data:

Variable	Description	Rows	Columns
x	Input data	Number of inputs	Number of patterns
t	Target data	Number of outputs	Number of patterns

An example of creating and saving a training set is:

```
x=[0:1:10];
t=2*x - 0.21*x.^2;
save data8 x t
```

This will create training data to train a network to approximate the function

$$y = 2x_1 + 0.21x_1^2$$

over the interval [0 10] and store in binary format in a file named data8.mat. The MATLAB script, bptrain, asks for the name of the file containing the training data, the number of hidden neurons, and the type of scaling to use. It also asks if the default error tolerance, maximum number of cycles, and initial learning rate is acceptable. If they are, the function backprop() is called and the network is trained. After training the weights, biases, and scaling parameters are saved in a file called weights.mat. The following is a diary of running the script bptrain on the training data defined above.

EDU» bptrain

Enter filename of input/output data vectors: data8

How many neurons in the hidden layer? 20

Input scaling method zscore=z, linear=l, none=n ?[z,l,n]:n

The default variables are:

Output error tolerence (RMS per output term) = .1
Maximum number of training cycles = 5000.
The initial learning rate is = 0.1.

Are you satisfied with these selections? [y,n]:y

This network has:

1 input neurons
20 neurons in the hidden layer
1 output neurons

There are 11 input/output pairs in this training set.

*** BP Training complete, error goal not met!
*** RMS = 1.780286e-001

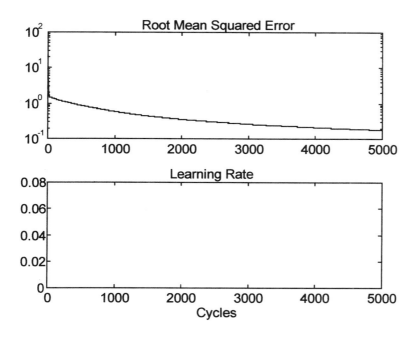

In this example, the error goal of 0.1 was not met in 8000 cycles; the final network error was 0.17. Note that this error is a root mean squared error rather than a sum of squared error. The RMS error is not dependent on the number of outputs and the number of training patterns. It is therefore more intuitive. It can be thought of as an average error per output rather than the total error over all outputs and training patterns.

After we train a network we usually want to test it. To test the network, we define a test set of input/output patterns of the function that was used to train the network. Since some of these patterns were not used to train the network, this checks for generalization. Generalization is the ability of a network to give the correct output for an input that was not in the training set. Networks are not expected to generalize outside of the training space and no confidence should be given to outputs generated from data outside of the training data.

```
load weights8
x=[0:.1:10];
t=2*x - 0.21*x.^2;
output = W2*[ones(size(x));logistic(W1*[ones(size(x));x])];
plot(x,t,x,output)
title('Function Approximation Verification')
xlabel('Input')
ylabel('Output')
```

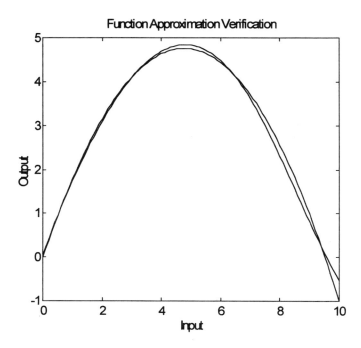

The network does a good job of learning the functional relationship between the inputs and outputs. It also generalizes well. Training longer to a lower error goal would improve the network performance.

8.6 Backpropagation Example

In this example a neural network is trained to recognize the letters of the alphabet. The first 16 letters (A through P) are defined on a 5x7 template and are each stored as a vector of length 35 in the data file letters.mat. The data file contains an input matrix x (size = 35 by 16) and a target matrix t (size = 4 by 16). The entries in a column of the t matrix are a binary coding of the letter of the alphabet in the corresponding column of x. The columns in the x matrix are the indices of the filled in boxes of the 5x7 template. A function letgph() displays the letter on the template. For example, to plot the letter A which is the first letter of the alphabet and identified by t=[0 0 0 0], we load the data file and plot the first column of x.

```
load('letters')
t(:,1)'
letgph(x(:,1));

ans =
     0     0     0     0
```

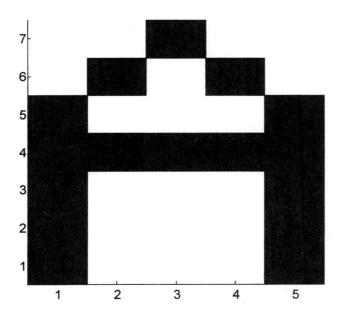

We will now train a neural network to identify a letter. The network has 35 inputs corresponding to the boxes in the template and has 4 outputs corresponding to the binary code of the letter. Since the outputs are binary, we can use a logistic output layer. The function bprop2 has a logistic/logistic architecture. We will use a beginning learning rate of 0.5, train to a maximum of 5000 cycles and a root mean squared error goal of 0.05. This may take a few minutes.

```
load letters
W1=0.5*randn(10,36);    % Initialize output layer weight matrix.
W2=0.5*randn(4,11);     % Initialize hidden layer weight matrix.
[W1 W2 RMS]=bprop2(x,t,W1,W2,.05,.1,5000);
semilogy(RMS);
title('Backpropagation Training');
xlabel('Cycles');
ylabel('Root Mean Squared Error')
```

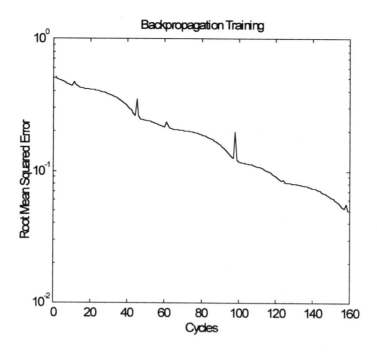

The trained network is saved in a file weights_1. We can now use the trained network to identify an input letter. For example, presenting the network with several letters resulted in:

```
load weight_1;
output0 = logistic(W2*[1;logistic(W1*[1;x(:,1)])])'
output1 = logistic(W2*[1;logistic(W1*[1;x(:,2)])])'
output7 = logistic(W2*[1;logistic(W1*[1;x(:,8)])])'
output12 = logistic(W2*[1;logistic(W1*[1;x(:,13)])])'

output0 =
    0.0188    0.0595    0.0054    0.0443
output1 =
    0.0239    0.0433    0.0607    0.9812
output7 =
    0.0590    0.9872    0.9620    0.9536
output12 =
    0.9775    0.9619    0.0531    0.0555
```

The network correctly identifies each of the test cases. You can verify this by comparing each output with its binary equivalent. For example, output7 should be [0 1 1 1], which is very close to its actual output. One may want to make the target outputs be in the range of [.1 .9] because outputs of 0 and 1.0 are not obtainable and may force the weights to very large values.

A neural network's ability to generalize is found in its ability to give a correct response to an input that was not in the training set. For example, a noisy input of an A may look like:

```
a=[0 0 1 0 0 0 0 1 0 1 0 0 0 1 1 1 1 1 1 1 0 0 0 1 1 0 0 0 1 1 0 0 1 1] ';
```

```
letgph(a);
```

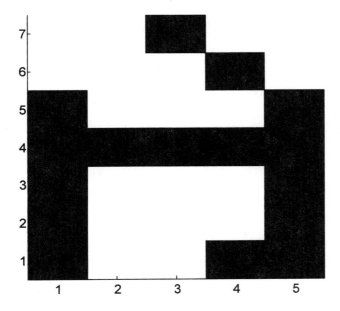

Presenting the network with this input results in:

```
OutputA = logistic(W2*[1;logistic(W1*[1;a])])'

OutputA =
    0.0165    0.1135    0.0297    0.0161
```

This output is very close to the binary pattern [0 0 0 0] which designates an 'A'. This show the networks ability to generalize, and more specifically, its tolerance to noise.

Chapter 9 Competitive, Associative and Other Special Neural Networks

9.1 Hebbian Learning

Recall from Section 7.3 that Hebb's training rule states:

> *"When the synaptic input and the neuron output are both active, the strength of the connection between the input and the output is enhanced."*

There are several methods of implementing a Hebbian learning rule, a supervised form is used in the perceptron learning rule of Chapter 7. This chapter explores the implementation of unsupervised learning rules and begins with an implementation of Hebb's rule. An unsupervised learning rule is one in which no target outputs are given.

If the output of the single layer network is active when the input is active, the weight connecting the two active nodes is enhanced. This allows the network to associate relationships between inputs and outputs, hence the name *associative* networks. The most simple unsupervised Hebb rule is:

$$w^{new} = w^{old} + \beta xy$$

Where: w_{AB} is the weight connecting input A to output B.
        β is the learning constant
        x is the input
        y is the output

The constant β controls the rate at which the network learns. If β is made large, few presentations are needed to learn an association and if β is made small, many presentations are needed.

If the weights between active neurons are only allowed to be enhanced, as in the equation above, there is no limit to their magnitude. Therefore, a rule that allows both learning and forgetting should be implemented. Stephen Grossberg [1982] states the weight change law as:

$$w^{new} = w^{old}(1-\alpha) + \beta xy$$

In this equation, α is the forgetting constant and controls the rate at which the memory of old information is allowed to decay away or be forgotten. Using this Hebbian update rule, the network constantly forgets old information and continually learns new information. The values of β and α control the speed of learning and forgetting and are usually set in the interval [0 1]. The update rule can be rewritten as:

$$\Delta \mathbf{W} = -\alpha \mathbf{W} + \beta \mathbf{xy}^T$$

This rule limits the magnitude of the weights to a value determined by α and β. Solving the above equation for $\Delta \mathbf{W}=0$, we find the maximum weight to be β/α when x and y are active. For example: if the learning rate β is set to 0.9 and there is a small forgetting rate $\alpha=0.1$, then the maximum weight is $(\beta/\alpha)*xy^T=9xy^T$.

Suppose a four input network, that has its weights initialized to the identity matrix, is presented with an input vector x.

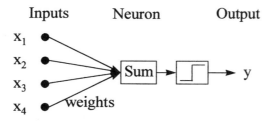

```
w=[1 0 0 0];              % Weight vector.
x=[1 0 0 1]';             % Input vector.
```

The inputs will be presented to the network and the weights will be updated with the unsupervised Hebbian learning rule with β = 0.1 and α =0.1.

```
a=0.1;                    % Forgetting factor.
b=0.1;                    % Learning factor.
yout=thresh(w*x-eps)      % Output.
del_w=-a*w+b*x'*yout;     % Weight update.
w=w+del_w                 % New weight.

yout =
    1
w =
    1.0000         0         0    0.1000
```

This rule is implemented in a function called hebbian(x,w,a,b,cycles). This function is called with cycles equal to the number of training iterations.

```
w=hebbian(x,w,a,b,100);   % Train the network with a Hebbian learning
                          % rule for 100 cycles.
w                         % Trained weight vector.

w =
    1.0000         0         0    1.0000
```

We can see that the weights are bounded by $(\beta/\alpha)*xy^T=1$. The network has learned to associate an input of [1 0 0 1] with an active output. Now if a degraded version of the input vector is input to the network, the network's output will still be active. For example, if the input is [0 0 0 1], the output will still be high. The network has learned to associate an x_1 or x_2 input with an active output.

9.2 Instar Learning

The Hebbian learning rule described above will continuously forget previous information due to the weight decay structure. A more useful structure would allow the network to forget only when it is learning. Since the network only learns when the output is active (y = 1), the network should only forget when the output is active. If we make the learning and forgetting rates equal, this Hebbian learning rule is called the Instar learning rule.

$$w_{AB}^{ne} = w_{AB}^{old}(1-\alpha y^T) + \beta xy^T = w_{AB}^{old} + \alpha(x - w_{AB}^{old})y^T$$

Rearranging:

$$w_{AB}^{ne} = w_{AB}^{old}(1-\alpha y^T) + \alpha x_A y^T$$

The first term allows the network to forget when the output is active and the second term allows the network to learn. This implementation causes the weight matrix to move towards new inputs. If the weight vector and input vector are normalized, this weight update rule is graphically represented as:

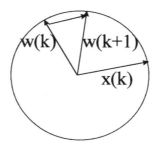

Instar Learning

The value of α determines the rate at which the weight vector is drawn towards new input vectors. The output of an Instar is the dot product of the weight vector and the input vector. Therefore, the Instar has the ability to learn an input vector and the output is a value corresponding to the degree that the input matches the weight vector. The Instar learns patterns and classifies patterns.

The following is an example of how the Instar network can learn to remember and recognize an input vector. Note that the weight vector and input vector are normalized and that the dot product must be >0.9 for the identification of the input vector to be positive.

```
w=rand(1,4)              % Initialize weight vector randomly.
x=[1 0 0 1]';            % Input vector to be learned.
x=x/norm(x);             % Normalize input vector.
w=w/norm(w);             % Normalize weight vector.
yout1=thresh(w*x-.9)     % Initial output.
a=0.8;                   % Learning and forgetting factor.
w=instar(x,w,a,10);      % Train for 10 cycles.
w                        % Final weight vector.
yout2=thresh(w*x-.9)     % Final output.

w =
    0.2190    0.0470    0.6789    0.6793
yout1 =
    0
w =
```

```
         0.7071    0.0000    0.0000    0.7071
yout2 =
     1
```

The network was able to learn a weight vector that would identify the desired input vector. A 0.9 criteria was used for identification.

9.3 Outstar Learning

The Instar network learns to identify input vectors and its dual network, called an Outstar network, can store and recall a vector. Putting these two network architectures together results in an associative memory.

The Outstar network also uses a Hebbian learning rule. Again, when the input and output are both active, the weight connecting the two is increased. The Outstar network is trained by applying an active signal at the input while applying the vector to be stored at the output. This is a type of supervised network since the desired output is applied to the output. After training, when the input is made active, the output will be the stored vector.

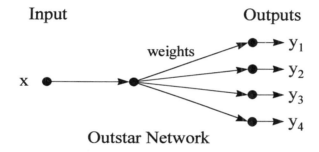

Outstar Network

For example, a network can be trained to store a vector v = [0 1 1 0]. After training, when the input is a 1, the output will be the learned vector; if the input is a 0, the output will be a 0. The Outstar uses a Hebbian learning rule similar to that of the Instar. If the learning and forgetting terms are equal, the Outstar learning rule for this simple recall network is:

$$\Delta w = -\alpha w x + \alpha v x = \alpha(w - v)x$$

The Outstar function, outstar(v,x,w,a), is used where:
    v is the vector to be learned
    x is the input vector
    w is the weight matrix of size(outputs,inputs)
    a is the learning rate.

```
v = [0 1 1 0];           % Vector to be learned.
x=1;                     % Input.
w=rand(1,4)              % Initialize weights randomly.
a=0.9;                   % Learning rate.
w=outstar(v,x,w,a,10);   % Train the network for 10 cycles.
yout=w*x                 % The trained network output.
```

```
w =
    0.9347    0.3835    0.5194    0.8310
yout =
    0.0000    1.0000    1.0000    0.0000
```

This algorithm can generalize to higher dimensional inputs. The general Outstar architecture is now:

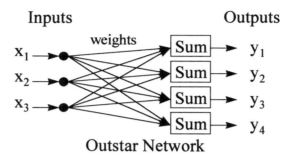

Outstar Network

If we want a network to output a certain vector depending on the active input node we will have a weight matrix versus a weight vector. Suppose we want to learn three vectors of four terms.

```
v1=[0 -1 0 0]';          % First vector.
v2=[1 0 -1 0]';          % Second vector.
v3=[0 1 0 0]';           % Third vector.
v=[v1 v2 v3];
w=rand(4,3);             % Random initial weight matrix.
w=outstar(v,x,w,a,10);   % Train the network for 10 cycles.
x=[1 0 0
0 1 0
0 0 1];                  % Three input vectors.
yout=w*x                 % Three output vectors.

yout =
    0.0000    1.0000    0.0000
   -1.0000    0.0000    1.0000
    0.0000   -1.0000    0.0000
    0.0000    0.0000    0.0000
```

The network learned to recall the correct vector for each of the three inputs. Although this implementation uses a supervised paradigm, the implementation could be presented in an unsupervised form. The unsupervised form uses an initial identity weight matrix, a learning rate equal to one, and only one pass of training. This method embeds the input matrix into the weight matrix in one pass but may not be useful when there are several patterns in the data set that are noisy. The one pass method may not generalize well from noisy or incomplete data.

9.4 Crossbar Structure

An associative network architecture with a crossbar structure is termed a bi-directional associative memory (BAM) by its developer, Bart Kosko [1988]. This methodology is really a matrix solution to an associative memory problem. The BAM does not undergo training as do most neural network architectures.

As an example of its implementation, consider three vector pairs (a,b).

```
a1=[1 -1 -1 -1 -1 1]';
a2=[-1 1 -1 -1 1 -1]';
a3=[-1 -1 1 -1 -1 1]';
b1=[-1 1 -1]';
b2=[1 -1 -1]';
b3=[-1 -1 1]';
```

Each vector a is associated with a vector b, and either vector can be the input or the output. Note that the terms of the vectors in a BAM must be ± 1. Three correlation matrices are formed by multiplying the vectors using the equation:

$$M_1 = A_1 B_1^T$$

```
m1=a1*b1';
m2=a2*b2';
m3=a3*b3';    % Three correlation matrices.
```

The three correlation matrices are then added to get a master weight matrix:

$$M = M_1 + M_2 + M_3$$

```
m=m1+m2+m3    % Master weight matrix.

m =
    -1     3    -1
     3    -1    -1
    -1    -1     3
     1     1     1
     3    -1    -1
    -3     1     1
```

The master weight matrix can now be used to get the vector associated with any input vector. This matrix can perform transformations in either direction. The resulting vector must be limited to the [1 -1] range. This is done using the signum() function.

$$A_i = MB_i \quad \text{or} \quad B_i = M^T A_i$$

For example:

```
A1=signum(m*b1)    % Recall a1 from b1.
B2=signum(m'*a2)   % Recall b2 from a2.

A1 =
     1
```

$$B2 = \begin{matrix} -1 \\ -1 \\ -1 \\ -1 \\ 1 \\ 1 \\ -1 \\ -1 \end{matrix}$$

We can see that the BAM was able to recall the associations stored in the master matrix memory. A discussion of the capacity and efficiency of the BAM network is given in the text.

9.5 Competitive Networks

Artificial neural networks that use competitive learning have only one output node activated at a time. The output nodes compete to be the one that is active, this is sometimes called a winner-take-all algorithm.

$$y = \sum_l w_{il} x_l = \mathbf{w}_i \mathbf{x}$$

where: i = output node index
l = input node index

The weights between the input and output nodes (w_{il}) are initially chosen as small random values and are continuously normalized. When training commences, the input vectors (**x**) search out which weight vector ($\mathbf{w}_i^*$) is closest to it.

$$\mathbf{w}_i^* \cdot \mathbf{x} \geq \mathbf{w}_i \cdot \mathbf{x}$$

The closest weight vector is then updated to make it closer to the input vector. The amount that the weight vector is changed is determined by the learning rate η.

$$\Delta w_{ij}^* = \eta \left(\frac{x_j}{\sum_j x_j} - w_{ij} \right)$$

Training involves the repetitive application of input vectors to the network in a random order or in turn. The weights are continually updated with each application. This process could continue changing the weight vectors forever; therefore, the learning rate η is reduced as training progresses until it eventually is negligible and the weight changes cease. This results in weight vectors (+) centered in clusters of input vectors (*) as shown in the following figure.

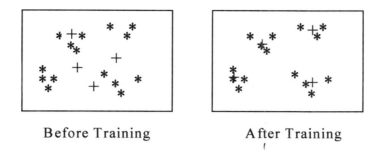

Before Training   After Training

Competitive Network

Generally, competitive learning networks are single-layered but there are several variants that are multi-layered. Some are briefly described by Hertz, Krogh, and Palmer [1991] but none will be discussed here. As is intuitively obvious, competitive networks perform clustering. They extract similarities from the input vectors and group or categorize them into specific clusters. These similar input vectors fire the same output node. They find uses in vector quantization problems such as data compression.

9.5.1 Competitive Network Implementation

A competitive network is a single layer network which only has one output activate at a time. This output corresponds to the neuron whose weight vector is closest to the input vector.

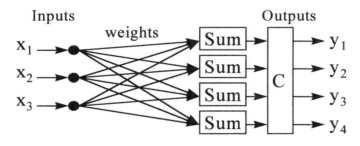

Competitive Network

Suppose we have three weight vectors already stored in a competitive network. The output of the network to an input vector is found with the following.

```
w=[1 1; 1 -1; -1 1];      % Cluster centers.
x=[1;3];                  % Input to be classified.
y=compete(x,w);           % Classify.
clg
plot(w(:,1),w(:,2),'*')
hold on
plot(x(1),x(2),'+')
hold on
plot(w(find(y==1),1),w(find(y==1),2),'o')
title('Competitive Network Clustering')
```

```
xlabel('Input 1');ylabel('Input2')
axis([-1.5 1.5 -1.5 3.5])
```

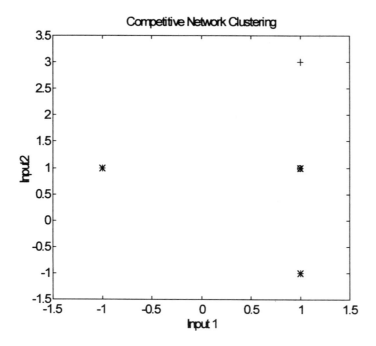

The above figure displays the cluster centers (*), shows the input vector (+), and shows that cluster 1 won the competition (o). Cluster center 1 is circled and is closest to the input vector labeled +. The next example shows how a competitive network is trained. It uses the instar learning rule to learn the cluster centers and since only one output is active at a time, only the winning weight vector is updated during each presentation.

Suppose there are a 11 input vectors (dimension=2) that we want to group into 3 clusters. The weight matrix is randomly initialized and the network is trained for 20 presentation of the 11 inputs.

```
x=[-1 5;-1.2 6;-1 5.5;3 1;4 2;9.5 3.3;-1.1 5;9 2.7;8 3.7;5 1.1;5 1.2]';
clg
plot(x(1,:),x(2,:),'*');title('Training Data'),xlabel('Input 1'),
ylabel('Input 2');
```

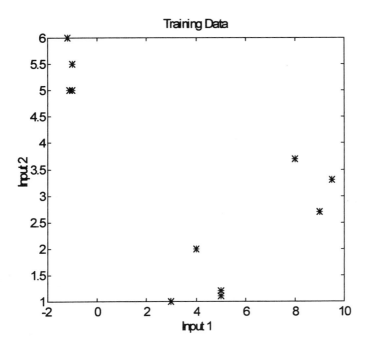

This data is sent to a competitive network for training. It is specified *a priori* that the network will have 3 clusters.

```
w=rand(3,2);
a=0.8;                   % Learning and forgetting factor.
w=trn_cmpt(x,w,a,20);    % Train for 20 cycles.
plot(x(1,:),x(2,:),'k*');title('Training Data'),xlabel('Input 1'),
ylabel('Input 2');hold on
plot(w(:,1),w(:,2),'o');hold off
```

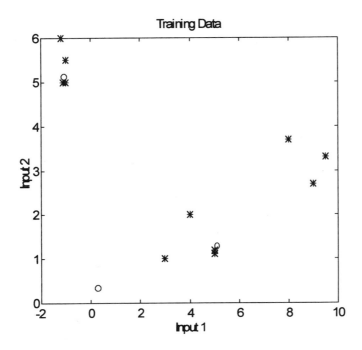

101

Note that the first weight vector never moves towards the group of data centered around (10,3). This neuron is called a "dead neuron" since its output is never activated. One method of dealing with dead neurons is to somehow give them an extra chance of winning a competition (to give them increased bias towards winning). To implement this increased bias, we will add a bias to all of the competitive neurons. The bias will be increased when the neurons doesn't win and decreased when the neuron does win. The function competb() will evaluate a competitive network with a bias.

```
dimension=2;                    % Input space dimension.
clusters=3;                     % Number of clusters to identify.
w=rand(clusters,dimension);     % Initialize weights.
b=.1*ones(clusters,1);          % Initialize biases.
a=0.8;                          % Learning and forgetting factor.
cycles=5;                       % Iteratively train for 5 cycles.
w=trn_cptb(x,w,b,a,cycles);     % Train network.
clg
plot(x(1,:),x(2,:),'*');title('Training Data'),xlabel('Input 1'),
ylabel('Input 2');hold on
plot(w(:,1),w(:,2),'o');hold off
```

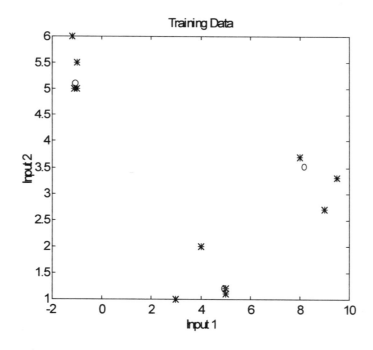

We can see that the dead neuron has come alive. It has centered itself in third cluster.

One detail that has not been discussed in detail is the selection of the learning rate. A large learning rate allows the network to learn fast but reduces its stability and leads to oscillatory behavior. An adaptive learning rate can be used that allows the network to initially learn fast and then slower as training progresses.

9.5 2 Self Organizing Feature Maps

The self-organizing feature map, or Kohonen network [1984], maps a high dimension input vector into a smaller dimensional pattern; usually this pattern is of dimension one or two. The conventional two dimensional feature map architecture is shown below.

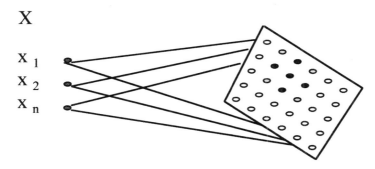

Self Organizing Feature Map

In a feature map, the geometrical arrangement or location of the outputs contains information about the input vectors. If the input vectors X_1 and X_2 are fairly similar, their outputs should be located close together; and if X_1 and X_2 are quite similar, then their outputs should be equal. This relationship can be realized by one of several different learning algorithms. It can be realized by using ordinary competitive learning with lateral connections weights in the output layer that excite nearby nodes and inhibit nodes that are farther away. It can also be realized by ordinary competitive learning where the weights of nearby neighbors are allowed to update along with the weights of the winning node. This realization is termed Kohonen's algorithm. Kohonen's algorithm clusters data in a way that preserves the topology of the inputs, thus geometrically revealing the similarities of the inputs. This similarity is defined as it was in competitive learning as the Euclidean distance from each other.

The Kohonen learning algorithm is:

$$\Delta w = \alpha(x - w_i) \quad \text{for } w_i \text{ near } x.$$

The weight updates are only performed for the winning neuron and its neighbors (w_i near x_i). For a 4x4, 2-dimensional feature map, the neurons nearest to the winning neuron are also updated. These neurons are the gray shaded neurons in the above figure. There may be a reduced update for neurons as a function of their distance from the winning neuron. This type of function is commonly referred as a "Mexican hat" function.

`mexhat;`

Mexican Hat Function

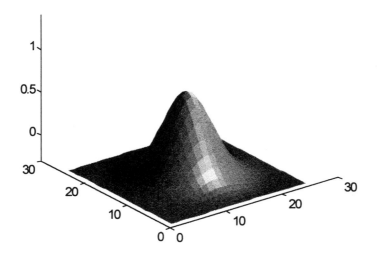

Suppose we have a two input network that is to organize the data into a one dimensional feature map of length 5.

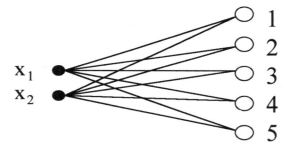

One Dimensional Feature Map

In the following example, we will organize 18 input pairs (this may take a long time).

```
x=[1 2;8 9;7 8;6 6;2 3;7 7;2 2;5 4;3 3;8 7;4 4; 7 6;1 3;4 5;8 8;5 5;6 7;9 9]';
plot(x(1,:),x(2,:),'*');title('Training Data'),xlabel('Input 1'),
ylabel('Input 2');
b=.1*ones(5,1);         % Small initial biases.
w=rand(5,2);            % Initial random weights.
tp=[0.7 20];            % Learning rate and maximum training iterations.
[w,b]=kohonen(x,w,b,tp);% Train the self-organizing map.
ind=zeros(1,18);
for j=1:18
    y=compete(x(:,j),w);
    ind(j)=find(y==1);
end
```

[x;ind]

```
ans =
  Columns 1 through 12
     1    8    7    6    2    7    2    5    3    8    4    7
     2    9    8    6    3    7    2    4    3    7    4    6
     1    5    4    3    1    4    1    2    1    4    2    4
  Columns 13 through 18
     1    4    8    5    6    9
     3    5    8    5    7    9
     1    2    5    3    4    5
```

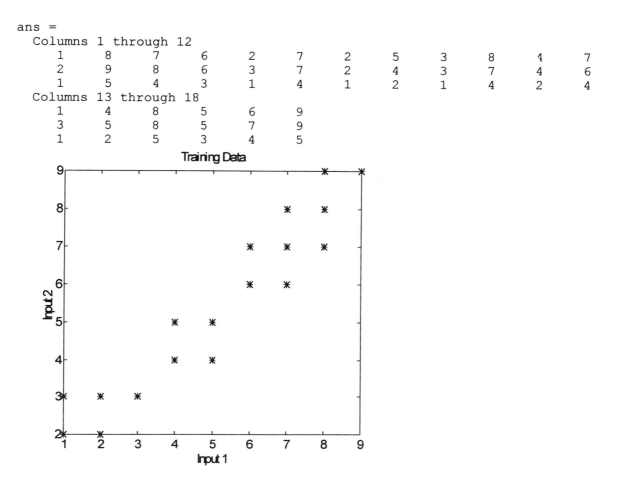

To observe the order of the classifications, we will use different symbols to plot the classifications.

```
clg
plot(w(:,1),w(:,2),'+')
hold on
plot(w(:,1),w(:,2),'-')
hold on
index=find(ind==1);
plot(x(1,index),x(2,index),'*')
hold on
index=find(ind==2);
plot(x(1,index),x(2,index),'o')
hold on
index=find(ind==3);
plot(x(1,index),x(2,index),'*')
hold on
index=find(ind==4);
plot(x(1,index),x(2,index),'o')
hold on
index=find(ind==5);
```

```
plot(x(1,index),x(2,index),'*')
hold on
axis([0 10 0 10]);
xlabel('Input 1'), ylabel('Input 2');
title('Self Organizing Map Output');
hold off
```

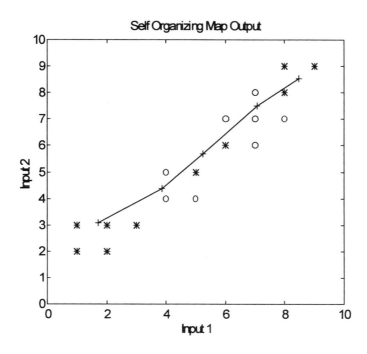

It is apparent that the network not only clustered the data, but it also organized the data so that the data near each other were put in clusters next to each other. The use of self organizing maps preserves the topography of the input vectors.

9.6 Probabilistic Neural Networks

The Probabilistic Neural Network (PNN) is a Bayesian Classifier put into a neural network architecture. This network is well described in *Fuzzy and Neural Approaches in Engineering*, by Lefteri H. Tsoukalas and Robert E. Uhrig. Timothy Masters has written two books that give good discussions of PNNs: *Practical Neural Network Recipes in C++* [1993a] and *Advanced Algorithms for Neural Networks* [1993b]. Both of these books contain disks with C++ version of a PNN code.

A PNN is a classifier. Although it can be used as a function approximator, this task is better performed by other iteratively trained neural network architectures or by the Generalized Regression Neural Network of Section 9.8.

One of the most common classifiers is the Nearest Neighbor classifier. This classifier classifies a pattern to be in the same class as its nearest neighbor, or more generally, its k nearest neighbors. A drawback of the Nearest Neighbor classifier is that sometimes the nearest neighbor may be an outlier from another class. In the figure below, the input marked as a ? would be classified as a o, even though it is in the middle of many x's. This is a weakness of the nearest neighbor classifier.

Classification Problem

The principal advantage of PNNs over other NN architectures is its speed of learning. Its weights are not trained through an iterative process, they are stored during what is commonly called the learning process. A second advantage is that the PNN has a solid theoretical foundation for making confidence estimates.

There are several major disadvantages to the PNN. First, the PNN must store all training patterns. This requires large amounts of memory. Secondly, during recall, all the training patterns must be processed. This requires a lengthy recall period. Also, the PNN requires a large representative training set for proper operation. Lastly, the PNN requires the proper choice of a width parameter called sigma. There are routines to choose this parameter in an optimal manner, but this is an iterative and sometimes lengthy procedure (see Masters 1993a). The width parameter may be different for each population (class), but the implementation presented here will use a single width parameter.

In summary, the Probabilistic Neural Network should be used only for classification problems where there is a representative training set. It can be trained quickly but has slow recall and is memory intensive. It has solid underlying theory and can produce confidence intervals. This network is simply a Bayesian Classifier put into a neural network architecture. The estimator of the probability density function uses the gaussian weighting function:

$$g(x) = \frac{1}{n}\sum_{i=1}^{n} e^{-\frac{\|x-x_i\|^2}{2\sigma^2}}$$

Where:
  n is the number of cases in a class
  x_i is a specific case in a class
  x is the input
  σ is the width parameter

This formula simply estimates the probability density function (PDF) as an average of separate multivariate normal distributions. This function is used to calculate the probability density function for each class. For example, if the training data consisted of three classes with populations of 23, 12, and 18. The above formula would be used to estimate the PDF for each of the three classes with n=23, 12 and 18.

A simple PNN will now be implemented that classifies an input as one of two classes. The training data consists of two classes of data with four vectors in each class. A test data point will be used to verify correct operation.

```
x=[-3 -2;-3 -3;-2 -2;-2 -3;3 2;3 3;2 2;2 3]; % Training data
y=[1 1 1 1 2 2 2 2]';         % Classifications of training data
xtest=[-.5 1.5];              % Vector to be classified.
plot(x(:,1),x(:,2),'*');hold;plot(xtest(1),xtest(2),'o');
title('Probabilistic Neural Network Data')
axis([-4 4 -4 4]); xlabel('Input1');ylabel('Input2');hold off;
```

```
Current plot held
```

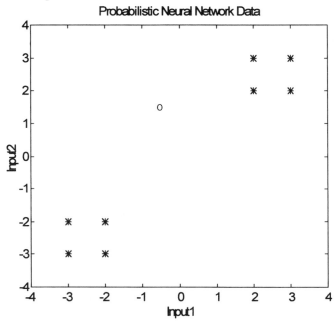

The test data point can be classified by a PNN.

```
a=3;                          % a is the width parameter: sigma.
classes=2;                    % x has two classifications.
[class,prob]=pnn(x,y,classes,xtest,a)  % Classify the test input: testx.

class =
    2
prob =
    0.1025    0.3114
```

This function properly classified the input vector as class 2 and output a measure of membership (0.3114). As a final example, we will use an input vector x=[-2.5 -2.5].

```
x=[-3 -2;-3 -3;-2 -2;-2 -3;3 2;3 3;2 2;2 3];   % Training data
y=[1 1 1 1 2 2 2 2]';                          % Classifications of training data
xtest=[-2.5 -2.5];                             % Vector to be classified.
plot(x(:,1),x(:,2),'*');hold;plot(xtest(1),xtest(2),'o');
title('Probabilistic Neural Network Data')
axis([-4 4 -4 4]); xlabel('Input1');ylabel('Input2');hold off
a=3;                                           % a is the width parameter: sigma.
classes=2;                                     % x has two classifications.
[class,prob]=pnn(x,y,classes,xtest,a)          % Classify the test input.

Current plot held
class =
     1
prob =
    0.9460    0.0037
```

The PNN properly classified the input vector to class 1. The PNN also outputs a number related to the membership of the input to each class (0.946 0.004). These numbers can be used as confidence values for the classification.

9.7 Radial Basis Function Networks

A Radial Basis Function Network (RBF) has been proven to be a universal function approximator [Park and Sandberg 1991]. Therefore, it can perform similar function mappings as a MLP but its architecture and functionality are very different. We will first examine the RBF architecture and then examine the differences between it and the MLP that arise from this architecture.

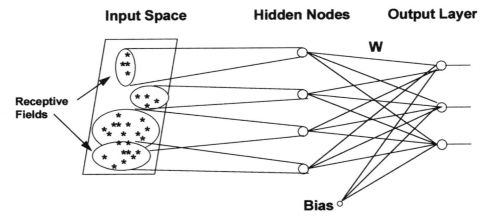

Radial Basis Function Network

A RBF network is a two layer network that has different types of neurons in the hidden layer and the output layer. The hidden layer, which corresponds to a MLP hidden layer, is a non-linear, local mapping. This layer contains radial basis function neurons which most commonly use a gaussian activation function (g(x)). These functions are centered over receptive fields. Receptive fields are areas in the input space which activate the local radial basis neurons.

$$g_j(x) = \exp\left[-(x - \mu_j)^2 / \sigma_j^2\right]$$

Where:
  x is the input vector.
  μ_j is the center of a region called a receptive field.
  σ_j is the width of the receptive field.
  $g_j(x)$ is the output of the jth neuron.

The output layer is a layer of standard linear neurons and performs a linear transformation of the hidden node outputs. This layer is equivalent to a linear output layer in a MLP, but the weights are usually solved for using a least square algorithm rather trained for using backpropagation. The output layer may, or may not, contain biases; the examples in this supplement do not use biases.

Receptive fields center on areas of the input space where input vectors lie, and serve to cluster similar input vectors. If an input vector (x) lies near the center of a receptive field (μ), then that hidden node will be activated. If an input vector lies between two receptive field centers, but inside the receptive field width (σ) then the hidden nodes will both be partially activated. When input vectors that lie far from all receptive fields there is no hidden layer activation and the RBF output is equal to the output layer bias values.

A RBF is a local network that is trained in a supervised manner. This contrasts with a MLP network that is a global network. The distinction between local and global is the made though the extent of input surface covered by the function approximation. An MLP performs a global mapping, meaning all inputs cause an output, while an RBF performs a local mapping, meaning only inputs near a receptive field produce an activation.

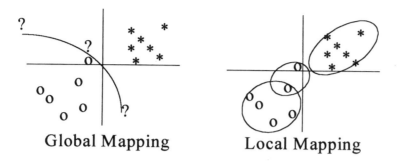

Global Mapping   Local Mapping

The ability to recognize whether an input is near the training set or if it is in an untrained region of the input space gives the RBF a significant benefit over the standard MLP. It can give a "don't know" output. Since networks generalize improperly and arbitrarily when operating in regions outside the training area, no confidence should be given to their outputs in those regions. When using an MLP, one cannot judge whether or not the input vector comes from these untrained regions; and therefore, one cannot judge whether the output contains significant information. On the other hand, an RBF can tell the user if the network is operating outside its training region and the user will know when to disregard the output. This ability makes the RBF the network of choice for safety critical applications or for applications that have a high financial impact.

The radial basis function y=gaussian(x,w,a) is given above and is implemented in an m-file where
    x is the input vector
    w is center of the receptive field
    a is the width of the receptive field
    y is the output value

```
x=[-3:.1:3]';        % Input space.
y=gaussian(x,0,1);   % Radial basis function centered at 0
plot(x,y);           % with a width of 1.
grid;
xlabel('input');ylabel('output')
title('Radial Basis Neuron')
```

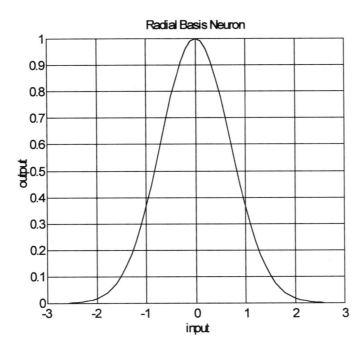

From the above figure, we can see that the output of the Gaussian function, evaluated at the width parameter, equals about one half. We also note that the function has an approximate zero output at distances of 2.5 times the width parameter. This figure shows the range of coverage of the gaussian activation function.

Designing an RBF neural network requires the selection of the radial basis function width parameter. This decision is not required for an MLP. The width should be chosen so that the receptive fields overlap but so that one function does not cover the entire input space. This means that several radial basis neurons have some activation to each input but all radial basis neurons are not highly active for a single input.

Another choice to be made is the number of radial basis neurons. Depending on the training algorithm used to implement the RBF, this may, or may not, be a decision made by the designer. For example, the MATLAB Neural Network Toolbox has two training algorithms. The first algorithm centers a radial basis neuron on each input vector. This leads to an extremely large network for input data composed of many patterns. The second algorithm incrementally adds radial basis neurons to reduce the training error to the preset goal.

There are several network architectures that will meet a specified error criteria. These architectures consist of different combinations of the radial basis function widths and the number of radial basis functions in the network. The following figure roughly shows the allowable combinations that may solve an example problem.

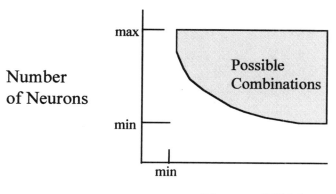

The maximum number of neurons is the number of input patterns, the minimum is related to the error tolerance and the complexity of the mapping. This minimum must be experimentally determined. A more complex map and a smaller tolerance requires more neurons. The minimum width constant should overlap the input patterns and the maximum should not cover the entire input space. Excessively large widths can sometimes give good results for data with no noise, but these systems usually fail under real world conditions in which noise exists. The reason that the system can train well with noise free cases is that a linear method is used to solve for the second layer weights. The use of a regression method will minimize the error, but usually at the expense of large weights and significant overfitting. This overfitting is apparent when there is noise in the system. A smaller width will do a better job of alerting that an input vector is outside the training space, while a larger width may result in a network of smaller size and faster execution.

9.7.1 Radial Basis Function Example

As an example of the implementation of a RBF network, a function approximation over an interval will be used.

```
x=[-10:1:10]';                  % Inputs.
y=.05*x.^3-.2*x.^2-3*x+20;      % Target outputs.
plot(x,y);
xlabel('Input');
ylabel('Output');
title('Function to be Approximated')
```

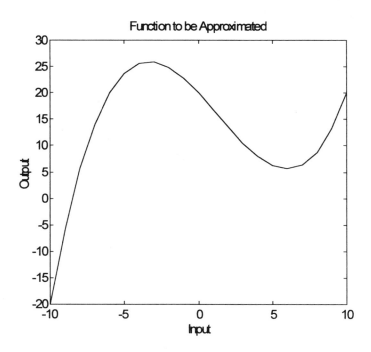

A radial basis function width of 4 will be used and the centers will be placed at [-8 -5 -2 0 2 5 8]. Most training routines will have an algorithm for determining the placement of the radial basis neurons, but for this example, they will simply be placed to cover the input space.

```
width=6;                    % Radial basis function width.
w1=[-8 -5 -2 0 2 5 8];      % Center of the receptive fields.
num_w1=length(w1);          % The number of receptive fields.
a1=gaussian(x,w1,width);    % Hidden layer outputs.
w2=inv(a1'*a1)*a1'*y;       % Pseudo inverse to solve for output weights.
```

The network can now be tested both outside the training region and inside the training region.

```
test_x=[-15:.2:15]';                              % Test inputs.
y_target=.05*test_x.^3-.2*test_x.^2-3*test_x+20;  % Test outputs.
test_a1=gaussian(test_x,w1,width);                % Hidden layer outputs.
yout=test_a1*w2;                                  % Network outputs.
plot(test_x,[y_target yout]);
title('Testing the RBF Network'); xlabel('Input'); ylabel('Output')
```

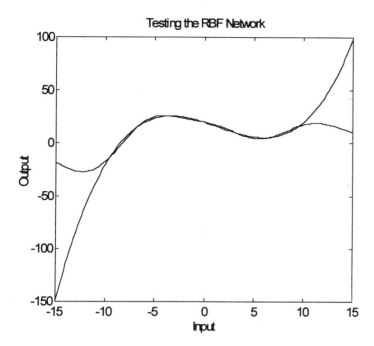

The network generalizes very well in the training region but poorly outside the training region. As the inputs get far from the training region, the radial basis neurons are not active. This would alert the operator that the network is trying to operate outside the training space and that no confidence should be given to the output value.

To show the tradeoffs between the size of the neuron width and the number of neurons in the network, we will investigate two other cases. In the first case, the width will be made so small that there is no overlap and the number of neurons will be equal to the number of inputs. In the second case, the spread constant will be made very large and some noise will be added to the data.

9.7.2 Small Neuron Width Example

The radial basis function width will be set to 0.2 so that there is no overlap between the neurons.

```
x=[-10:1:10]';                  % Inputs.
y=.05*x.^3-.2*x.^2-3*x+20;      % Target outputs.
width=.2;                       % Radial basis function width.
w1=x';                          % Center of the receptive fields.
a1=gaussian(x,w1,width);        % Hidden layer outputs.
w2=inv(a1'*a1)*a1'*y;           % Solve for output weights.
test_x=[-15:.2:15]';                            % Test inputs.
y_target=.05*test_x.^3-.2*test_x.^2-3*test_x+20;   % Test outputs.
test_a1=gaussian(test_x,w1,width);              % Hidden layer outputs.
yout=test_a1*w2;                                % Network outputs.
plot(test_x,[y_target yout]);
title('Testing the RBF Network');xlabel('Input');ylabel('Output')
```

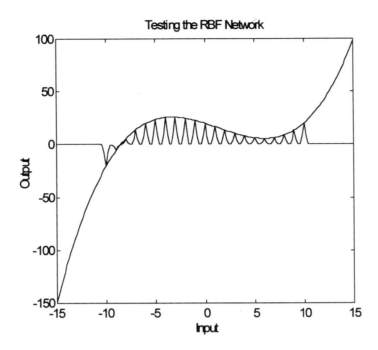

The above figure shows that the width parameter is too small and that there is poor generalization inside the training space. For proper overlap, the width parameter needs to be at least equal to the distance between input patterns.

9.7.3 Large Neuron Width Example

A very large radial basis function width equal to 200 will be used. Such a large width parameter causes each radial basis function to cover the entire input space. When this occurs, the radial basis functions are all highly activated for each input value. Therefore, the network may have problems learning the desired mapping.

```
width=200;                   % Radial basis function width.
w1=[-8 -3 3 8];              % Center of the receptive fields.
a1=gaussian(x,w1,width);     % Hidden layer outputs.
w2=inv(a1'*a1)*a1'*y;        % Solve for output weights.
test_x=[-15:.2:15]';                        % Test inputs.
y_target=.05*test_x.^3-.2*test_x.^2-3*test_x+20;   % Test outputs.
test_a1=gaussian(test_x,w1,width);          % Hidden layer outputs.
yout=test_a1*w2;                            % Network outputs.
plot(test_x,[y_target yout]);
title('Testing the RBF Network');
xlabel('Input');
ylabel('Output')
```

```
Warning: Matrix is close to singular or badly scaled.
         Results may be inaccurate. RCOND = 3.273238e-017
```

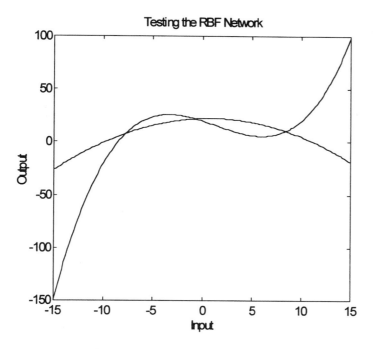

The hidden layer activations ranged from 0.9919 to 1.0. This made the regression solution of the output weight matrix very difficult and ill-conditioned. The use of such a large width parameter causes numerical problems and also makes it difficult to know when an input vector is outside the training space.

9.8 Generalized Regression Neural Network

The Generalized Regression Neural Network [Specht 1991] is a feedforward neural network best suited to function approximation tasks such as system modeling and prediction. Although it can be used for pattern classification, the Probabilistic Neural Network discussed in Section 9.6 is better suited to those applications.

The GRNN is composed of four layers. The first layer is the input layer and is fully connected to the pattern layer. The second layer is the pattern layer and has one neuron for each input pattern. This layer performs the same function as the first layer RFB neurons: its output is a measure of the distance the input is from the stored patterns. The third layer is the summation layer and is composed of two types of neurons: S-summation neurons and a single D-summation neuron (division). The S-summation neuron computes the sum of the weighted outputs of the pattern layer while the D-summation neuron computes the sum of the unweighted outputs of the pattern neurons. There is one S-summation neuron for each output neuron and a single D-summation neuron. The last layer is the output layer and divides the output of each S-summation neuron by the output of the D-summation neuron. A general diagram of a GRNN is shown below.

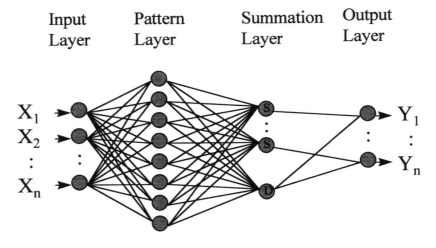

Generalized Regression Neural Network

The output of a GRNN is the conditional mean given by:

$$\hat{\mathbf{Y}} = \frac{\sum_{j=1}^{T} \mathbf{W}^T \exp\left(-\frac{D_t^2}{2\sigma^2}\right)}{\sum_{j=1}^{T} \exp\left(-\frac{D_t^2}{2\sigma^2}\right)}$$

Where the exponential function is a Gaussian function with a width constant sigma. Note that the calculation of the Gaussian is performed in the pattern layer, the multiplication of the weight vector and summations are performed in the summation layer, and the division is performed in the output layer.

The GRNN learning phase is similar to that of a PNN. It does not learn iteratively as do most ANNs; but instead, it learns by storing each input pattern in the pattern layer and calculating the weights in the summation layer. The equations for the weight calculations are given below.

The pattern layer weights are set to the input patterns.

$$\mathbf{W}_p = \mathbf{X}^T$$

The summation layer weights matrix is set using the training target outputs. Specifically, the matrix is the target output values appended with a vector of ones that connect the pattern layer to the D-summation neuron.

$$\mathbf{W}_s = [\mathbf{Y} \ \text{ones}]$$

To demonstrate the operation of the GRNN we will use the same example used in the RBF section. The training patterns will be limited to five vectors distributed throughout the input space. A width parameter of 4 will also be used. The training parameters must cover the training space and the set should also contain the values at any minima or maxima. The input training vector is chosen to be [-10 -6.7 -3.3 0 3.3 6.7 10]. First we calculate the weight matrices.

```
x=[-10 -6.7 -3.3 0 3.3 6.7 10]';            % Training inputs.
y_target=.05*x.^3-.2*x.^2-3*x+20;           % Generate the target outputs.
[Wp,Ws]=grnn_trn(x,y_target)                % Calculate the weight matrices.
plot(x,y_target)
title('Training Data for a GRNN');
xlabel('Input');
ylabel('Output')

Wp =
   -10.0000   -6.7000   -3.3000        0    3.3000    6.7000   10.0000
Ws =
   -20.0000    1.0000
    16.0838    1.0000
    25.9252    1.0000
    20.0000    1.0000
     9.7189    1.0000
     5.9602    1.0000
    20.0000    1.0000
```

The GRNN will now be simulated for the training data.

```
x=[-10 -6.7 -3.3 0 3.3 6.7 10]';
a=2;
y=grnn_sim(x,Wp,Ws,a);
y_actual=.05*x.^3-.2*x.^2-3*x+20;
```

```
plot(x,y_actual,x,y,'*')
title('Generalization of a GRNN');
xlabel('Input');
ylabel('Output')
```

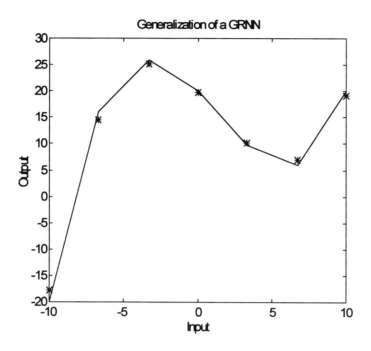

The recall performance for the network is very dependent on the width parameter. A small width parameter gives good recall of the training patterns but poor generalization. A larger width parameter would give better generalization but poorer recall. The choice of a good width parameter is necessary to having good performance. Usually, the largest width parameter that gives good recall is optimal. In the example above, a width parameter of 2.5 was found to be the maximum width that has good recall.

Next we check for correct generalization by simulating the GRNN over the trained region. To show the effects of the width parameter, the GRNN is simulated with a width parameter being too small (a = .5), too large (a = 5), and optimal (a = 2).

```
x=[-10:.5:10]';
a=.5;
y=grnn_sim(x,Wp,Ws,a);
y_actual=.05*x.^3-.2*x.^2-3*x+20;
plot(x,y_actual,x,y,'*')
title('Generalization of GRNN: a = 0.5')
xlabel('Input');
ylabel('Output'  )
```

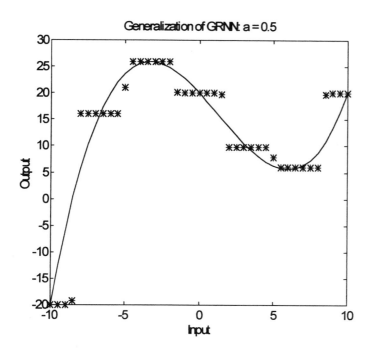

```
x=[-10:.5:10]';a=5;
y=grnn_sim(x,Wp,Ws,a);
y_actual=.05*x.^3-.2*x.^2-3*x+20;
plot(x,y_actual,x,y,'*')
title('Generalization of GRNN, a = 5')
xlabel('Input');ylabel('Output')
```

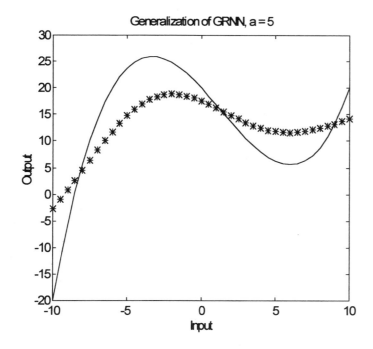

```
x=[-10:.5:10]';a=2;
y=grnn_sim(x,Wp,Ws,a);
```

```
y_actual=.05*x.^3-.2*x.^2-3*x+20;
plot(x,y_actual,x,y,'*')
title('Generalization of GRNN: a = 2.5')
xlabel('Input');ylabel('Output')
```

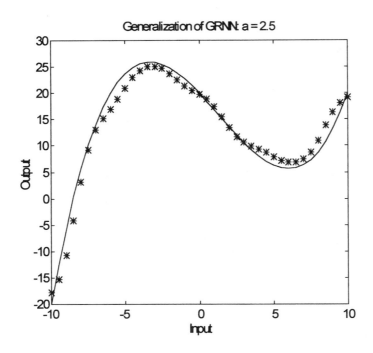

Note that with the proper choice of training data and width parameter, the network was able to generalize with very few training parameters. If there is nothing known about the function, a large training set must be chosen to guarantee it is representative. This would make the network very large (many pattern nodes) and would require much memory and long recall times. Clustering techniques can be used to select a representative training set, thus reducing the number of pattern nodes.

Chapter 10 Dynamic Neural Networks and Control Systems

10.1 Introduction

Dynamic neural networks require some sort of memory. This memory allows the network to exhibit temporal behavior; behavior that is not only dependent on present inputs, but also on prior inputs. There are two major classes of dynamic networks: Recurrent Neural Networks (RNN) and Time Delayed Neural Networks (TDNN). Recurrent Neural Networks are networks with internal time delayed feedback connections. The two most common RNN designs are the Elman network [Elman 1990] and the Jordan network [Jordan 1986]. In an Elman network, the hidden layer outputs are fed back through a one step delay to dummy input nodes. The Elman network can learn temporal patterns as well as spatial patterns because it can store information. The Jordan network is a recurrent architecture similar to the Elman network but it feeds back the output layer rather than the hidden layer. Recurrent Neural Networks are difficult to train due to the feedback connections. Usual methods are Real Time Recurrent Learning (RTRL) [Williams and Zipser, 1989], and Back Propagation Through Time (BPTT) [Werbos 1990].

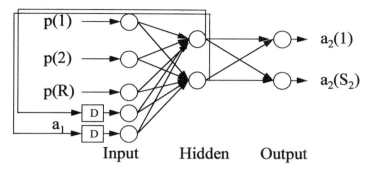

Elman Recurrent Neural Network

Time Delay Neural Networks (TDNNs) can learn temporal behavior by using not only the present inputs, but also past inputs. TDNNs accomplish this by simply delaying the input signal. The neural network architecture is usually a standard MLP but it can also be a RBF, PNN, GRNN, or other feedforward network architecture. Since the TDNN has no feedback terms, it is easily trained with standard algorithms.

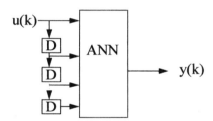

Time Delay Neural Network Model

Specific applications using these dynamic network architectures will be discussed in later sections of this chapter.

10.2 Linear System Theory

The educational version of MATLAB provides many functions for linear system analysis. This section will provide simple examples of the usage of some of these functions. Theoretical issues and derivation will not be discussed in this supplement. We will examine the Fast Fourier Transform (fft) and the Power Spectral Density (psd) functions.

Suppose you have a periodic signal that is 256 time steps long and is a combination of sine waves of 3 different frequencies. Taking the fft of that signal results in

```
t=[1:1:512];                                      % Time
f1=.06*(2*pi); f2=.1*(2*pi); f3=.4*(2*pi);        % Three frequencies
sig=2*sin(f1*t)+sin(f2*t)+1.5*sin(f3*t);          % Periodic signal
plot(t,sig);title('Periodic Signal');
ylabel('Amplitude');xlabel('Time');axis([0 512 -5 5]);
```

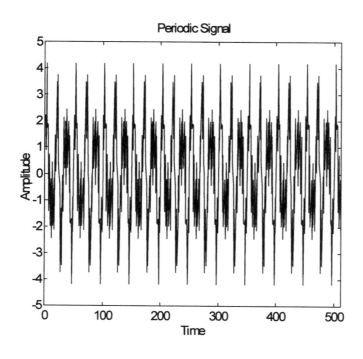

```
Y=fft(sig,256);          % Take Fast Fourier Transform
Pyy=Y.*conj(Y)/256;      % Find the normalized amplitude of the FFT.
f=[1:128]/256;           % Calculate a scale for the y axis.
plot(f,Pyy(1:128))       % Points 129:256 are symmetric.
xlabel('Frequency');ylabel('Power'); title('Power Spectral Density');
```

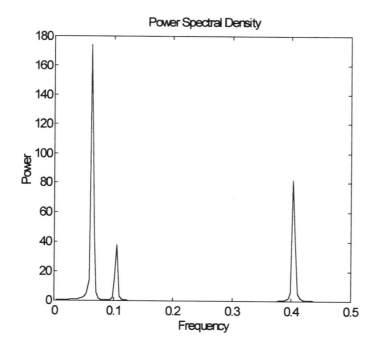

The plot of the power in each frequency band shows the three frequency components of the signal and the magnitudes of the signals. A noise component is usually inherent to most signals.

```
noisy_sig=sig+randn(size(sig));;  % Add normally distributed noise.
plot(t,noisy_sig);title('Noisy Periodic Signal');
ylabel('Amplitude');
xlabel('Time');
axis([0 512 -5 5]);
```

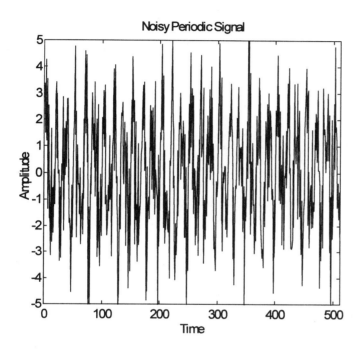

```
Y=fft(noisy_sig,256);     % Take Fast Fourier Transform
Pyy=Y.*conj(Y)/256;       % Find the normalized amplitude of the FFT.
f=[1:128]/256;            % Calculate a scale for the y axis.
plot(f,Pyy(1:128))        % Points 129:256 are symmetric.
xlabel('Frequency');
ylabel('Power');
title('Power Spectral Density');
```

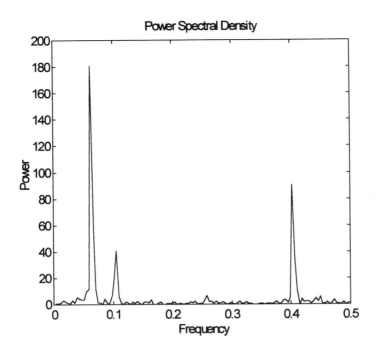

The figure shows the noise level rising across the entire spectrum. The PSD function can also be used to plot the Power Spectral Density with a dB scale.

```
psd(noisy_sig,256,1);    % Plots a Power Spectral Density [dB].
```

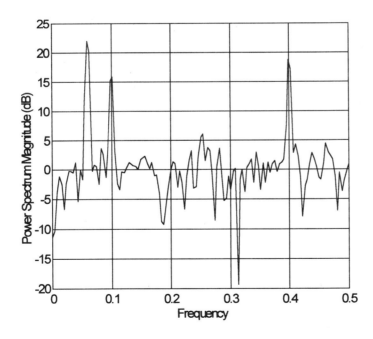

The use of these signal processing functions may be necessary when designing a neural or fuzzy system that uses inputs from the frequency domain.

10.3 Adaptive Signal Processing

Neural Networks can learn off-line or on-line. On-line learning neural networks are usually called adaptive networks because they can adapt to changes in the input or target signals. One example of an adaptive neural network is the single input adaptive transverse filter. This is a specific case of the TDNN with one layer of linear neurons.

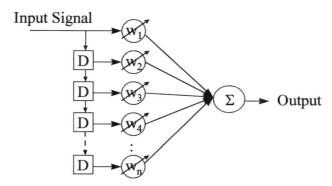

Single Input Adaptive Transverse Filter

Suppose that we have a Finite Duration Input Response (FIR) filter implemented by a TDNN (Shamma). FIR filters are sometimes used over Infinite Duration Input Response (IIR) filters such as Chebyshev, Elliptic, and Bessel due to their simplicity, stability, linearity, and finite response duration. The following example will show how a TDNN can be used as a FIR filter.

Suppose that a fairly noisy signal needs to be filtered so that it can be used as an input to a PID controller or other analog device. A FIR implemented as a neural network can be used for this application. Consider the following filtering example.

```
t=[250:1:400];
sig=sin(.01*t)+sin(.03*t)+.2*randn(1,151);
plot(t,sig);
title('Noisy Periodic Signal');
ylabel('Amplitude');
xlabel('Time');
```

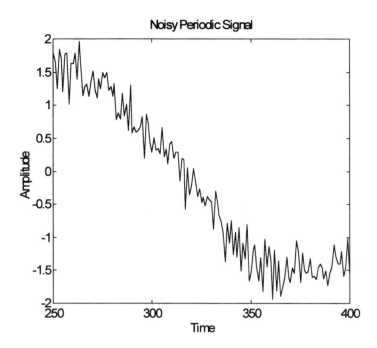

To simulate a TDNN we must construct an input matrix containing the delayed values of the input. For example, if we have an input vector of ten values that is being input to a TDNN with 3 delays, we call the function delay_in(x,d) with the number of delays d=3. This function returns an input vector with 4 rows since there are four inputs to the network. This function does not pad zeros into the inputs, so the input vector shrinks by 3 patterns. For example:

```
x=[0 1 2 3 4 5 6 7 8 9]';
xd=delay_in(x,3)

xd =
     0     1     2     3
     1     2     3     4
     2     3     4     5
     3     4     5     6
     4     5     6     7
     5     6     7     8
     6     7     8     9
```

Suppose we have a linear neural network FIR filter implemented with a TDNN with 5 delays. This filter can be used to process noisy data.

```
d=5;                                % Number of delays.
x=delay_in(sig',d);                 % Construct delayed input matrix.
w=[.2 .2 .2 .2 .1 .1]';             % Weight matrix.
y=x*w;                              % Calculate outputs.
td=t(d:length(sig)-1);
plot(td,sig(d:length(sig)-1),td,y');
title('Noisy Periodic Signal and Filtered Signal');
ylabel('Amplitude');
```

```
xlabel('Time');
```

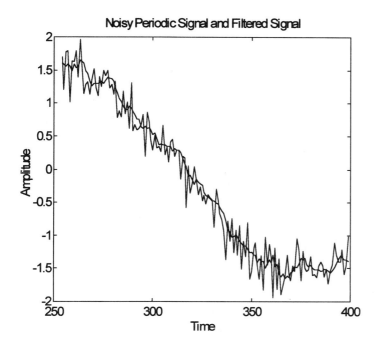

We can see that the neural network filter removed a large portion of the noise. It actually performs this operation by calculating a linear weighted average over a window. In this example, the window is six time steps long. This filter can be trained on line to give specific response characteristics.

10.4 Adaptive Processors and Neural Networks

The FIR example of the previous section is a dynamic network in the sense that is can model temporal behavior. This section demonstrates how a neural network can be dynamic in the sense that its parameters change with time. The example given trains a linear network on-line. This allows a network to adaptively learn non-stationary (time-varying) trends. The figure below shows a block diagram of an adaptive neural network used for system identification. The neural network can be a TDNN if temporal behavior is necessary or can be a static mapping without time delays. Systems whose characteristics (mathematical models) do not change with time can be modeled with a neural network that is trained off-line. The example of the section will only require a static mapping since the system output only depends on the current inputs but will be trained adaptively since one parameter is non-stationary.

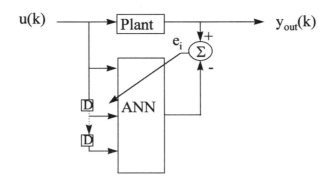

Parallel Identification Model

Let us consider a simple linear network used for adaptive system identification. In this example, the weights are the linear system coefficients of a non-stationary system. A linear neural network can be adaptively trained on-line to follow the non-stationary coefficient. Suppose the system is modeled by:

$$y = 2x_1 - 3x_2 - 1 + .01t$$

Initially, the offset term is -1 and this term changes to +1 as time reaches 200 seconds. A simple linear network with two inputs can be trained on-line to estimate this system. The two weighting coefficients will converge to 2 and -3 while the bias will start at -1 and move towards +1. The choice of the learning rate will affect the speed of learning and the stability of the network. A larger learning rate will allow the network to track faster but will reduce its stability while a lower learning rate will produce a stable system with slower tracking abilities.

```
W=[0 0 0];                      % Initialize weights and bias to zero.
Weights=[];                     % Store weight and bias values over time.
lr=.4;                          % Learning rate.
for i=1:200                     % Run for 200 seconds.
   x=rand(2,1);                 % System has random inputs.
   y=2*x(1)-3*x(2)-1+.01*i;     % Simulate system.
   [W]=adapt(x,y,W,lr);         % Train network.
   Weights=[Weights W'];        % Save weights and bias to plot.
end
plot(Weights')                  % Plot parameters.
title('Weights and Bias Approximation Over Time')
xlabel('Time');
ylabel('Weight and Bias Values')
text(50,2.5,'W1');
text(50,-2.5,'W2');
text(50,0,'B1');
```

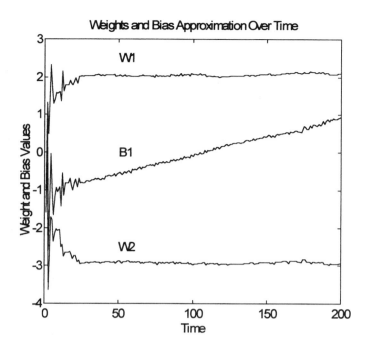

The figure shows that the network properly identified the system by about 30 seconds and the bias correctly tracked the non-stationary parameter over time. This type of learning paradigm could be used to train the FIR filter used in the previous section.

10.5 Neural Networks Control

There are user supplied MATLAB toolkits that implement neural network based system identification and control paradigms:

The NNSYSID Toolbox is located at: http://kalman.iau.dtu.dk/Projects/proj/nnsysid.html and the NNCTRL Toolkit is at: http://www.iau.dtu.dk/Projects/proj/nnctrl.html.

These toolboxes were developed by Magnus Morgaard of the Institute of Automation, Technical University of Denmark. The toolboxes and user guides: Technical Report 95-E-773 and Technical Report 95-E-830 can be downloaded free of charge.

For a more in depth discussion of the use of neural networks for system identification and control refer to *Advanced Control with MATLAB & SIMULINK* by Moscinski and Ogonowski. For further reading on the use of neural network for control see Irwin, Warwock and Hunt, 1995; Miller, Sutton and Werbos, 1990; Mills, Zomaya and Tade,1996; Narendra and Parthasarathy, 1990; Omatu, Khalid and Yusof, 1996; Pham and Liu, 1995; White and Sofga,1992;or Zbikowski and Hunt, 1996.

There are five general methods for implementing neural network controllers (Werbos pp. 59-65, in Miller Sutton and Werbos 1990):
1. Supervised Control
2. Direct Inverse Control

3. Neural Adaptive Control
4. Back-Propagation Through Time
5. Adaptive Critic Methods

10.5.1 Supervised Control

In supervised control, a neural network is trained to perform the same actions as another controller (mechanical or human) for given inputs and plant conditions. After the neural controller is trained, it replaces the controller.

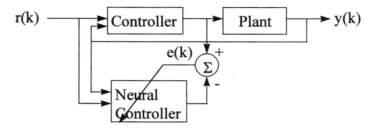

Neural Controller Training

10.5.2 Direct Inverse Control

In Direct Inverse Control, a Neural Network is trained to model the inverse of a plant. This modeling is similar to the system identification problem discussed in Section 10.6.

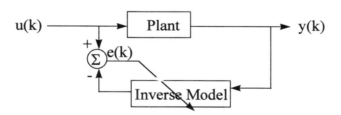

Inverse System Identification

After the neural network learns the inverse model, it is used as a forward controller. This methodology only works for a plant that can be modeled or approximated by an inverse function (F^{-1}). Since $F(F^{-1}) = 1$ the output (y(k)) approximates the input (u(k)).

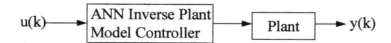

Direct Inverse Control

10.5.3 Model Referenced Adaptive Control

When the plant model changes with time due to wear, temperature affects, etc., neural adaptive control may be the best technique to use. Model Referenced Adaptive Control (MRAC) adapts the controller characteristics so that the controller/plant combination performs like a reference plant.

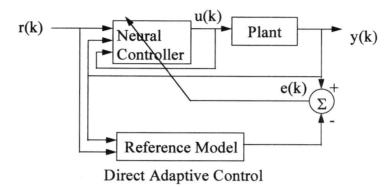

Direct Adaptive Control

Since the plant lies between the neural network and the error term, there is no method to directly adjust the controllers weights to reduce the error. Therefore, indirect control must be used.

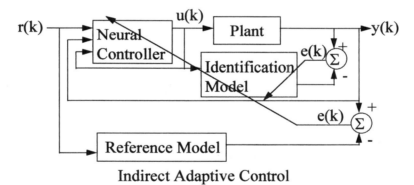

Indirect Adaptive Control

In indirect adaptive control, an ANN identification model is used to model the non-linear plant. If necessary, this model may be updated to track the plant. The error signals can now be backpropagated through the identification model to train the neural controller so that the plant response is equal to that of the reference model. This method uses two neural networks, one for system identification and one for MRAC.

10.5.4 Back Propagation Through Time

Back Propagation Through Time (BPTT) [Werbos 1990] can be used to move a system from one state to another state in a finite number of steps (if the system is controllable). First a system identification neural network model must be trained so that the error signals can be propagated through it to the controller, then the controller can be trained with a BPTT paradigm.

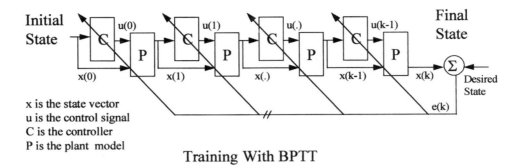

x is the state vector
u is the control signal
C is the controller
P is the plant model

Training With BPTT

BPTT training takes place in two steps:
1. The plant motion stage, where the plant takes k time steps.
2. The weight adjustment stage, where the controller's weights are adjusted to make the final state approach the target state.

It is important to note that there is only one set of weights to adjust because there is only one controller. Many iterations are run until performance is as desired.

10.5.5 Adaptive Critic

Often a decision has to be made without an exact conclusion as to its effectiveness (e.g. chess), but an approximation of its effectiveness can be obtained. This approximation can be used to change the control system. This type of learning is called reinforcement learning.

A critic evaluates the results of the control action: if it is good, the action is reinforced, if it is poor, the action is weakened. This is a trial and error method and uses active exploration when the gradient of the evaluation system in terms of the control action is not available. Note that this is an approximate method, and should only be used when a more exact method is unavailable.

A neural network based adaptive critic system uses one ANN to estimate the utility $J(k)$ of the state $x(k)$. This utility is a measure of the goodness of the state. It also uses a second ANN that trains with reinforcement learning to produce an input to the system ($u(k)$) that produces a good state, $x(k)$.

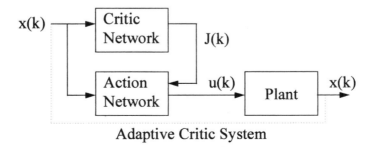

Adaptive Critic System

10.6 System Identification

The task of both conventional and neural network based system identification is to build a mathematical model of a dynamic system based on empirical data. In neural network based system identification, the internal weights and biases of the neural network are adjusted to make the model outputs similar to the measured outputs. Conventional methods use empirical data and regression techniques to estimate the coefficients of difference equations (ARX, ARMAX) or state space representations (see [Ljung 1987]).

10.6.1 ARX System Identification Model

A dynamic model is one whose output is dependent on past states of the system which may be dependent on past inputs and outputs. A static model's output at a specific time is only dependent on the input at that time. The basic conventional dynamic model is the difference equation and the most common difference equation form is the ARX model: Autoregressive with Exogenous Input Model which is sometimes called the Equation Error Model. This model uses past inputs and outputs to predict current outputs. For example, examine the following ARX model.

$$y(t) + a_1 y(t-1) + \ldots + a_{na} y(t-n_a) = b_1 u(t-1) + \ldots + b_{nb} u(t-n_b) + e(t)$$

The coefficients are represented by the set: $\theta = [a_1\ a_2\ \ldots\ a_{na}\ \ b_1\ b_2\ \ldots\ b_{nb}\ ]'$. The designer sets the structure and a regression technique is used to solve for the coefficients. This form can be visualized as

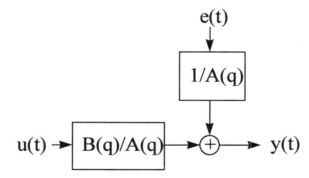

where:

$A(q) = 1 + a_1 q^{-1} \ldots a_{na} q^{-na}$
$B(q) = b_1 q^{-1} \ldots b_{nb} q^{-nb}$

and:

$y(t) = [B(q)/A(q)]u(t) + [1/A(q)]e(t)$.

This model assumes that the poles (roots of $A(q)$) are common between the dynamic model and the noise model. This model is a linear model, one of the major advantages of using neural networks is their non-linear approximating capabilities.

10.6.2 Basic Steps of System Identification

There are three phases of system identification:
A. Collect experimental input/output data.
B. Select and estimate candidate model structures.
C. Validate the models and select best model.

These phases can be accomplished by using the following general procedure:
1. Design an experiment and collect data. This data must be *Persistently Exciting*; meaning that the training set has to be representative of the entire class of inputs that may excite the system.
2. Process the data to filter and remove outliers, etc.
3. Select a model structure.
4. Compute the best parameters for that structure.
5. Examine the model's properties.
6. If the properties are acceptable quit, else goto 3.

10.6.3 Neural Network Model Structure

There are two basic neural network model structures: the parallel identification structure and the series parallel structure. The Parallel Identification Structure has direct feedback from the networks output to its input. It uses its estimate of the output to estimate future outputs. Because of this feedback, it has no guarantee of stability and requires dynamic backpropagation training. This structure should only be used if the actual plant outputs are not available.

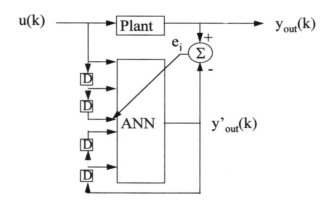

Parallel Identification Model

The Series-Parallel Identification Structure does not use feedback. Instead, it uses the actual plant output to estimate future system outputs. Therefore, static backpropagation training can be used and there are proofs to guarantee stability and convergence.

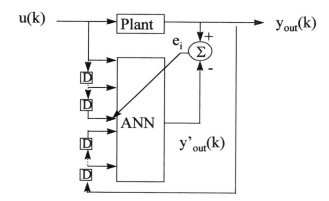

Series-Parallel Identification Model

Once the general model structure is chosen, the model order must be selected. The model order is the number of past signals to use as regressors. This can be estimated through knowledge of the system or through experimentation.

The network system identification model is then trained and tested. The error terms, also called residuals, should be white and independent of the input. Therefore, they are tested with an autocorrelation function and a crosscorrelation function with the inputs. Excessive correlation between the residuals and delayed inputs of outputs is evidence that the delayed inputs or outputs have information that can be used to reduce the estimation error.

If the output residuals have a high correlation with lagged input signals, there is reason to believe that the lagged input signal should be included as an input. This is one method that can be used to experimentally determine the number of lagged inputs to use for input to the model. For example, if a second order system is being identified with a neural network that only uses inputs delayed by one time step, the crosscorrelation between the two time step lagged input and the output will be large. This means that there is information in the two time step lagged input that can be used to reduce the estimation error.

As an example of an ARX model, consider:

$$y(t) - 1.5y(t-T) + 0.7y(t-2T) = 0.9u(t-2T) + 0.5u(t-3T) + e(t)$$

Here, the output at time t is dependent on the two past outputs, delayed inputs of 2 and 3 time steps and the disturbance. The basic steps in setting up a system identification problem are the same for a conventional model and a neural network based model. The ARX structure is defined by:

1. The number of delayed outputs to include ($y(t-T)$ $y(t-2T)$).

2. The time delay of the system. In this case the input does not effect the output for 2T.
3. The number of delayed inputs to use (u(t-2T) u(t-3T)).

A neural network model structure is defined by the same inputs but is also defined by the network architecture with includes the number and type of network hidden layers and hidden nodes.

10.6.4 Tank System Identification Example

This section presents an example that deals with the construction of a neural network system identification model for the tank system of Section 6.1. The function dy=tank_mod(t,y), is a non-linear model of the tank system where t is the simulation time, y(1) is the current state of tank (level), y(2) is the input signal (voltage to valve), and the output dy is the change in the tank state (change in level). To design a system identification model, input/output data must be collected for the operating range and input conditions of the tank. The operating state is the tank level and the input is the voltage supplied to the inlet valve. The state and input defined in the function tank_mod(t,y) cover the following ranges:

Y(1): Tank level (0-36 inches).
Y(2): Voltage being applied to the control valve (-4.5 to 1 volt).

To cover all possible operating conditions, we simulate the tank model over all combinations of the input and state.

```
x1=[0:3:36]';                                    % Level range.
x2=[-4.5 :.5:1]';                                % Input voltage range.
x=combine(x1,x2);                                % Input combinations.
dx=zeros(length(x1)*length(x2),1);               % Level changes vector.
for i=1:size(dx);                % Simulate the system for all combinations.
   dx(i)=tank_mod(1,x(i,:));
end
save tank_dat x dx                               % Save the simulation training data.
```

Now that we have training data, we can train a neural network to model the system. The inputs to the neural network model are the state: x(k) and the voltage going to the valve actuator: u(k). The output will be the change in the tank level: dx. By using a tank model with output dx and training an ANN with an input x(k) and u(k) we are able to avoid using a recurrent or time delay neural network.

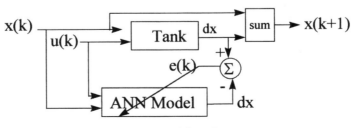

System Identification

The backpropagation training script is used to train a single layer neural network with 5 hidden neurons. The training data file has inputs x(k) and u(k) and the change in state dx is the target.

```
load tank_dat          % Tank inverse model simulation data.
t=dx';                 % Change in state.
x=x';                  % State and input voltage.
save tank_sys x t;     % Tank data in neural network training format.
```

The bptrian script trained the tank system identification model and the weights were saved in sys_wgt.mat.

```
load tank_sys  % Tank training data.
load sys_wgt   % Neural Network weights for system ID model.
[inputs,pats]=size(x);
output = linear(W2*[ones(1,pats);logistic(W1*[ones(1,pats);x])]);
subplot(2,1,1);
plot([t;output]');
title('System Identification Results')
ylabel('Actual and Estimated Output')
subplot(2,1,2);
plot(output-t);  % Calculate the error.
ylabel('Error');xlabel('Pattern Number');
```

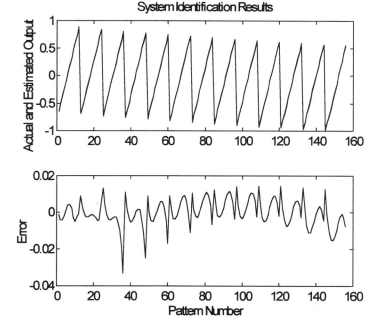

The top plot in the above figure shows that the neural network gives the correct output for the training set. The lower plot shows that the error levels are very small. The following is a comparison of outputs for a sinusoidal input. The time step must be the same as the integration time step of the analytical model. This checks for proper generalization.

```
t=[0:1:40];                          % Simulation time.
x(1)=15;                             % Initial tank level.
X(1)=15;                             % Initial tank estimator level.
u=sin(.2*t)-1.5;                     % Input voltage to control valve.
load sys_wgt              % Neural Network weights for system ID model.
[inputs,pats]=size(x);
for i=1:length(u);                   % Simulate the system.
   dx(i)=tank_mod(1,[x(i) u(i)]);    % Calculate change in state.
   estimate=linear(W2*[1;logistic(W1*[1;x(i);u(i)])]);
   x(i+1)=x(i)+dx(i);                % Update the actual state.
   X(i+1)=X(i)+estimate;             % Update state extimate.
end
plot(t,[x(1:length(u));X(1:length(u))]);
title('Simulation Testing of Tank Model')
xlabel('Time');ylabel('Tank Level');
```

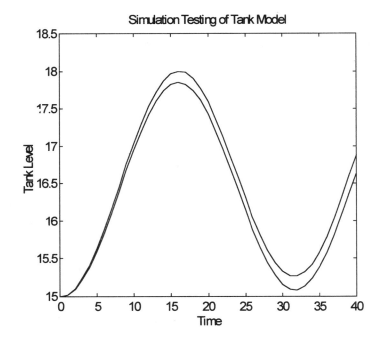

This simulation is being run in an open loop parallel-identification mode. That is, the value used for the current state input into the neural network is the neural network's estimate. This structure allows the errors to build up over time. When a closed loop structure is chosen, the estimate is much better.

```
t=[0:1:40];                          % Simulation time.
x(1)=15;                             % Initial tank level.
X(1)=15;                             % Initial tank estimator level.
u=sin(.2*t)-1.5;                     % Input voltage to control valve.
load sys_wgt              % Neural Network weights for system ID model.
[inputs,pats]=size(x);
```

```
for i=1:length(u);                    % Simulate the system.
   dx(i)=tank_mod(1,[x(i) u(i)]);     % Calculate change in state.
   estimate=linear(W2*[1;logistic(W1*[1;x(i);u(i)])]);
   x(i+1)=x(i)+dx(i);                 % Update the actual state.
   X(i+1)=x(i)+estimate;              % Update state estimate.
end
plot(t,[x(1:length(u));X(1:length(u))]);
title('Simulation Testing of Tank Model')
xlabel('Time');ylabel('Tank Level');
```

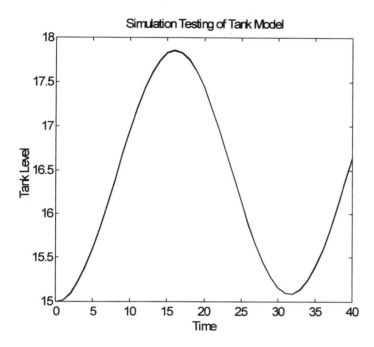

Since the actual state is used at each simulation time step, the errors are not allowed to build up and the neural network more closely tracks the actual system.

10.7. Implementation of Neural Control Systems

The example of this section deals with the use of an inverse neural network system identification model for direct inverse control of the tank problem of Section 6.1. The data constructed in Section 10.6.4 will be used to train an inverse system model. Again, the operating state is the tank level and the input is the voltage to the valve. The state and input are defined in the function tank_mod(t,y) cover the range:

Y(1): Tank level (0-36 inches).
Y(2): Voltage being applied to the control valve (-4.5 to 1 volt)

To construct a direct inverse neural network controller we need to train a neural network to model the inverse of the tank. The inputs to the neural network inverse model will be the state: x(k) and the desired change in state: dx. The output will be the input voltage going to the valve actuator: u(k). By using a tank model with output dx and training an

ANN with an input dx we are able to avoid using a recurrent or time delay neural network.

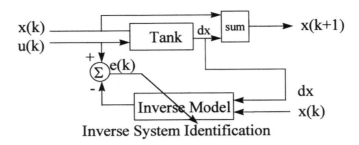

Inverse System Identification

The backpropagation training script was used to train a single layer neural network with 5 hidden neurons. We must set up the training data file correctly. The inputs x are the state x and the change in state dx. The target is the input u.

```
load tank_dat          % Tank inverse model simulation data.
t=x(:,2)';             % Valve actuator voltage.
x=[x(:,1) dx]';        % State and desired change in state.
save tank_trn x t;     % Tank data in neural network training format.
```

The bptrian script trained the inverse model and the weights were saved in tank_wgt.mat.

```
load tank_trn % Tank training data.
load tank_wgt % Neural Network weights for inverse system ID model.
[inputs,pats]=size(x);clg;
output = linear(W2*[ones(1,pats);logistic(W1*[ones(1,pats);x])]);
plot(output-t);  % Compare NN model results with tank analytical model.
title('Training Results');xlabel('Pattern Number');ylabel('Error')
```

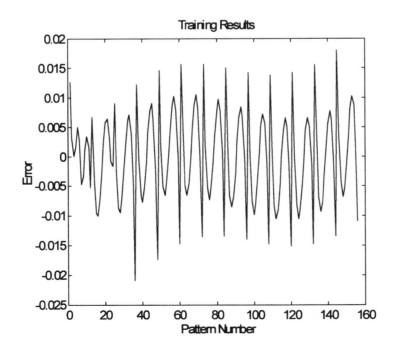

After the neural network is trained, it is put into the direct inverse control framework. The input to the inverse tank model controller is the current state and the desired state. The output of the controller is the voltage input to the valve actuator.

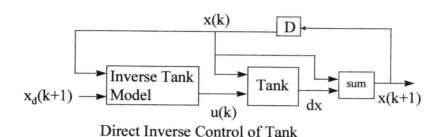

Direct Inverse Control of Tank

A simulation of the inverse controller can be run using tanksim(xinitial,x_desired). The output of this simulation is the level response of the neural network controlled tank. The desired level has been changed from 10 inches to 20 inches in this simulation.

```
tank_sim(10,20);
```

```
The tank and controller are simulated for 40 seconds, please be patient.
```

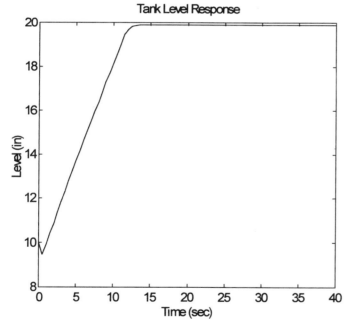

This controller has a faster speed of response and less steady state error than the fuzzy logic controller. This is shown in the next plot where both outputs are plotted together.

```
hold on
tankdemo(10,20);
text(2,18,'Neural Net');
text(18,18,'Fuzzy System');
```

```
hold off
```

The tank and controller are simulated for 40 seconds, please be patient.

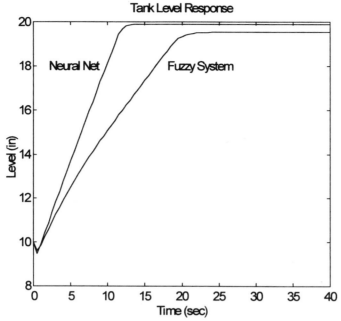

Chapter 11 Practical Aspects of Neural Networks

11.1 Neural Network Implementation Issues

There are several choices to be made when implementing neural networks to solve a problem. These choices involve the selection of the training and testing data, the network architecture, the training method, the data scaling method, and the error goal. Since over 90% of all neural network implementations use backpropagation trained multi-layer perceptrons, we will only discuss their implementation in this section. Sections 8.3 and 8.4 discussed scaling methods and weight initialization so those topics will not be revisited. The rest of these choices will be discussed in this chapter.

11.2 Overview of Neural Network Training Methodology

The figure below shows the methodology to follow when training a neural network. First you must collect or generate the data to be used for training and testing the neural network. Once this data is collected, it must be divided into a training set and a test set. The training set should cover the input space or should at least cover the space in which the network will be expected to operate. If there is not training data for certain conditions, the output of the network should not be trusted for those inputs. The division of the data into the training and test sets is somewhat of an art and somewhat of a trial and error procedure. You want to keep the training set small so that training is fast, but you also want to exercise the input space well which may require a large training set.

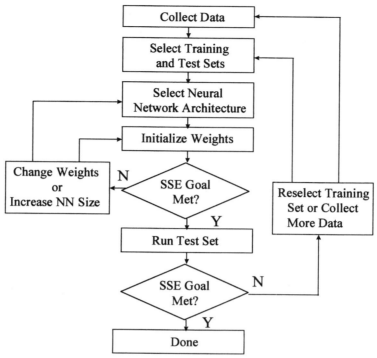

Neural Network Training Flow Chart

Once the training set is selected, you must choose the neural network architecture. There are two lines of thought here. Some designers choose to start with a fairly large network that is sure to have enough degrees of freedom (neurons in the hidden layer) to train to the desired error goal; then, once the network is trained, they try to shrink the network until the smallest network that trains remains. Other designers choose to start with a small network and grow it until the network trains and its error goal is met. We will use the second method which involves initially selecting a fairly small network architecture.

After the network architecture is chosen, the weights and biases are initialized and the network is trained. The network may not reach the error goal due to one or more of the following reasons.

1. The training gets stuck in a local minima.
2. The network does not have enough degrees of freedom to fit the desired input/output model.
3. There is not enough information in the training data to perform the desired mapping.

In case one, the weights and biases are reinitialized and training is restarted. In case two, additional hidden nodes or layers are added, and network training is restarted. Case three is usually not apparent unless all else fails. When attempting to train a neural network, you want to end up with the smallest network architecture that trains correctly (meets the error goal); if not, you may have overfitting. Overfitting is described in greater detail in Section 11.4.

Once the smallest network that trains to the desired error goal is found, it must be tested with the test data set. The test data set should also cover the operating region well. Testing the network involves presenting the test set to the network and calculating the error. If the error goal is met, training is complete. If the error goal is not met, there could be two causes:

1. Poor generalization due to an incomplete training set.
2. Overfitting due to an incomplete training set or too many degrees of freedom in the network architecture.

The cause of the poor test performance is rarely apparent without using crossvalidation checking which will be discussed in Section 11.4.3. If an incomplete test set is causing the poor performance, the test patterns that have high error levels should be added to the training set, a new test set should be chosen, and the network should be retrained. If there is not enough data left for training and testing, data may need to be collected again or be regenerated. These training decisions will now be covered in more detail and augmented with examples.

11.3 Training and Test Data Selection

Neural network training data should be selected to cover the entire region where the network is expected to operate. Usually a large amount of data is collected and a subset of that data is used to train the network. Another subset of that data is then used as test data to verify the correct generalization of the network. If the network does not generalize well on several data points, that data is added to the training data and the network is retrained. This process continues until the performance of the network is acceptable.

The training data should bound the operating region because a neural network's performance cannot be relied upon outside the operating region. This ability is called a network's extrapolation ability. The following is an example of a network that is being used outside of the region where it was trained.

```
x=[0:1:10];
t=2+3*x-.4*x.^2;
save data11 x t
```

The above code segment creates the training data used to train a network to approximate the following function:

$$f(x)=2+3*x-0.4*x.\wedge 2$$

```
load data11
plot(x,t)
title('Function Approximation Training Data')
xlabel('Input');ylabel('Output')
```

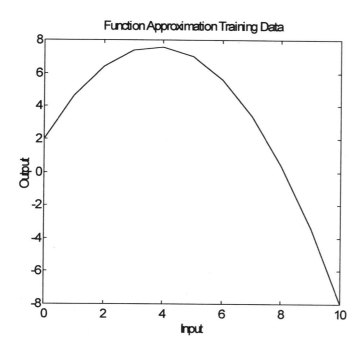

We will choose a single hidden layer network architecture with 2 logistic hidden neurons and train to either an average error level of 0.2 or 5000 epochs. The network was trained using bptrain and zscore scaling resulting in:

*** **BP Training complete, error goal not met!**
*** **RMS = 2.239272e-001**

After several tries, the network never reached the error goal in 5000 epochs. For this example, we will continue with the exercise and plot the training performance. Continued training may result in a network that meets the initial error criteria.

```
load weight11
subplot(2,1,1);
semilogy(RMS);
title('Backpropagation Training Results');
ylabel('Root Mean Squared Error')
subplot(2,1,2);
plot(LR)
ylabel('Learning Rate')
xlabel('Cycles');
```

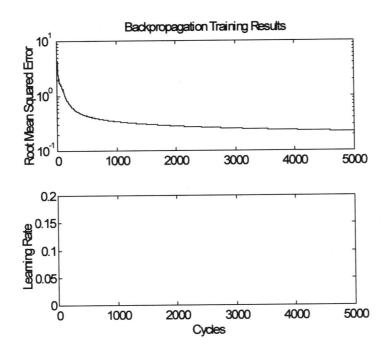

Now that the network is trained, we can check for generalization performance inside the training region. Remember that when scaling is used for training, those same scaling parameters must be used for any data input to the network.

```
x=[0:.01:10];y=2+3*x-0.4*x.^2;
load weight11;clg;X=zscore(x,xm,xs);
output = linear(W2*[ones(size(x));logistic(W1*[ones(size(x));X])]);
plot(x,y,x,output);title('Function Approximation Verification')
xlabel('Input');ylabel('Output')
```

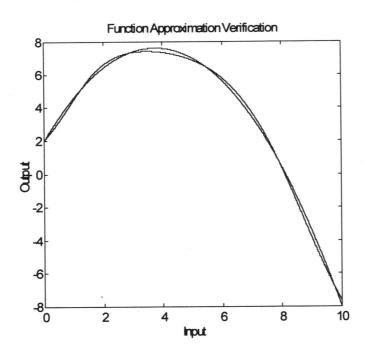

We can see that the network generalizes very well within the training region. Now we look at how the network extrapolates outside of the training region.

```
x=[-5:.1:15];
y=2+3*x-0.4*x.^2;
load weight11
X=zscore(x,xm,xs);
output = W2*[ones(size(x));logistic(W1*[ones(size(x));X])];
plot(x,y,x,output)
title('Function Approximation Extrapollation')
xlabel('Input')
ylabel('Output')
```

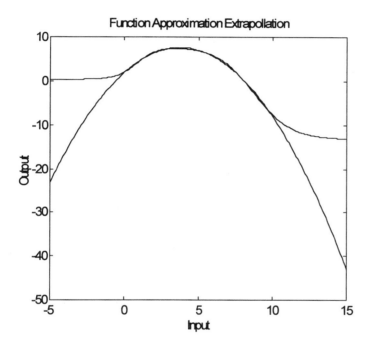

We see that the network generalizes very well within the training region (0-10) but poorly outside of the training region. This shows that a neural network should never be expected to operate correctly outside of the region where it was trained.

11.4 Overfitting

Several parameters affect the ability of a neural network to overfit the data. Overfitting is apparent when a networks error level for the training data is significantly better than the error level of the test data. When this happens, the data learned the peculiarities of the training data, such as noise, rather than the underlying functional relationship of the model to be learned. Overfitting can be reduced by:

1. Limiting the number of free parameters (neurons) to the minimum necessary.
2. Increasing the training set size so that the noise averages itself out.
3. Stopping training before overfitting occurs.

Three examples will now be used to illustrate these three methods. A robust training routine would use all of the above methods to reduce the chance of overfitting.

11.4.1 Neural Network Size.

In the example of Section 11.3, we see that the function can be approximated well with a network having 2 hidden neurons. Let us now train a network with data from a more realistic model. This model will have 20% noise added to simulate noisy data that would be measured from a process.

```
x=[0:1:10];
y=2+3*x-.4*x.^2;
randn('seed',3);    % Set seed to original seed.
t=2+3*x-.4*x.^2+0.2*randn(1,11).*(2+3*x-.4*x.^2);
save data12 x t
plot(x,y,x,t)
title('Training Data With Noise')
xlabel('Input');ylabel('Output')
```

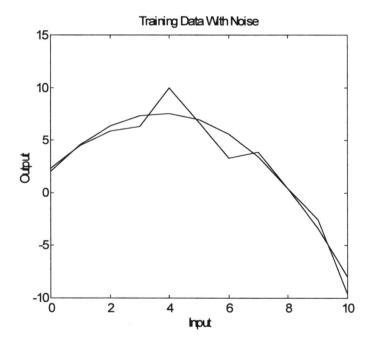

We trained a neural network with the same architecture as above (2 neurons) using the noisy data.

*** BP Training complete, error goal not met!
*** RMS = 1.276182e+000

The error only trained down to 1.27, but this is to be expected since we did not want the network to learn the noise.

```
x=[0:1:10];
y=2+3*x-0.4*x.^2;
load data12
load weight12
X=zscore(x,xm,xs);
output = linear(W2*[ones(size(x));logistic(W1*[ones(size(x));X])]);
clg
plot(x,y,x,output,x,t)
title('Function Approximation Verification')
xlabel('Input')
ylabel('Output')
```

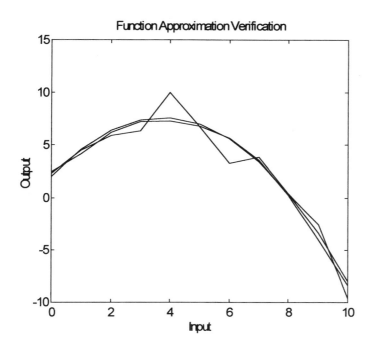

In the above figure, the smoothest line is the actual function, the choppy line is the noisy data and the line closest to the smooth line is the network approximation. We can see that the neural network approximation is smooth and follows the function very well. Next we increase the degrees of freedom to more than is necessary for approximating the function. This case will use a network with 4 hidden neurons.

*** **BP Training complete, error goal not met!**
*** **RMS = 4.709244e-001**

```
x=[0:1:10];
y=2+3*x-0.4*x.^2;
load data12
load weight13
X=zscore(x,xm,xs);
output = linear(W2*[ones(size(x));logistic(W1*[ones(size(x));X])]);
clg
plot(x,y,x,output,x,t)
title('Function Approximation Verification')
```

```
xlabel('Input')
ylabel('Output')
```

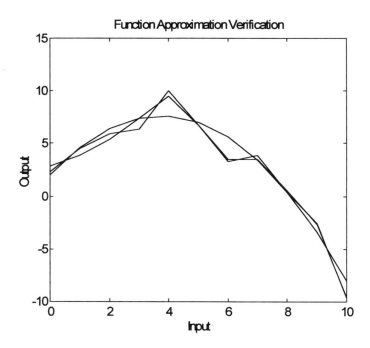

The above figure is the output of a network with 4 neurons trained with the noisy data. We can see that when extra free parameters are used in the network, the approximation is severely overfitted. Therefore, a network with the fewest number of free parameters that can be trained to the error goal should be used. This statement requires that a realistic error goal be set.

A function with 11 training points containing 20% random noise for outputs that average around 6, will have a RMS error goal of .2*6=1.2. This is about the value that we got in the properly trained example above. If we stop training of an overparameterized network when that error goal is met, we reduce the chances of overfitting the model. A neural network with 4 hidden neurons is now trained with an error goal of 1.2.

*** **BP Training complete after 151 epochs!** ***
*** **RMS = 1.194469e+000**

The network learned much faster (151 epochs versus 5000 epochs) than the network with only two neurons. Lets look at the generalization.

```
x=[0:1:10];
y=2+3*x-0.4*x.^2;
load data12
load weight14
X=zscore(x,xm,xs);
output = linear(W2*[ones(size(x));logistic(W1*[ones(size(x));X])]);
clg
```

```
plot(x,y,x,output,x,t)
title('Function Approximation Verification')
xlabel('Input')
ylabel('Output')
```

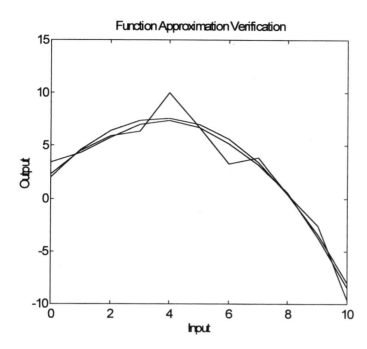

The network generalized much better than the overparameterized network with an unrealistically low error goal. This illustrates two methods that can be used to reduce the chance of overfitting.

When the actual signal is not known, a realistic error goal can be found by filtering the signal to smooth out the noise, then calculating the difference between the smoothed signal and the noisy signal. This difference is a rough approximation of the amount of noise in the signal.

11.4.2 Neural Network Noise

As discussed above, when there is noise in the training data, a method to calculate the RMS error goal needs to be used. If there is significant noise in the data, increasing the number of patterns in the training set can reduce the amount of overfitting. The following example will illustrate this point. In this example the test set size is increased to 51 patterns.

```
x=[0:.2:10];
randn('seed',100)
y=2+3*x-.4*x.^2;
t=2+3*x-.4*x.^2+0.2*randn(1,size(x,2)).*(2+3*x-.4*x.^2);
save data13 x t
plot(x,y,x,t)
title('Training Data With Noise')
```

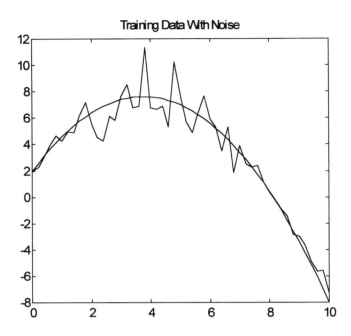

Using this data we will now train a network with 5 hidden neurons. The results are as follows.

*** BP Training complete, error goal not met!
*** RMS = 1.062932e+000

Seeing that the RMS error is less than 1.2, we can expect some overfitting.

```
x=[0:.2:10];
y=2+3*x-0.4*x.^2;
load data13
load weight15
X=zscore(x,xm,xs);
output = linear(W2*[ones(size(x));logistic(W1*[ones(size(x));X])]);
clg
plot(x,y,x,output,x,t)
title('Function Approximation Verification With 5 Neurons')
xlabel('Input')
ylabel('Output')
```

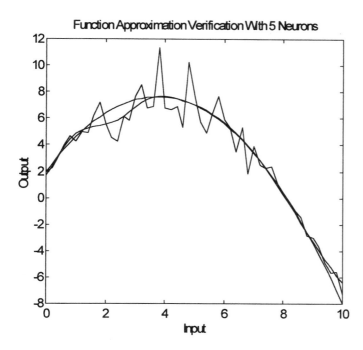

The above figure shows that the network generalized very well even though there were too many free parameters in the model. This shows that using a more representative training set tends to average out the noise in the signals without having to stop training at an appropriate error goal or having to estimate the number of hidden neurons. The network does sag at the input=2 region, to avoid this, use more samples or reduce the number of hidden neurons.

11.4.3 Stopping Criteria and Cross Validation Training

The last method of reducing the chance of overfitting is cross validation training. Cross validation training uses the principle of checking for overfitting during training. This methodology uses two sets of data during training. One set is used for training and the other is used to check for overfitting. Since overfitting occurs when the neural network models the training data better than it would other data, checking data is used during training to test for this overlearning behavior.

At each training epoch, the RMS error is calculated for both the test set and the checking set. If the network has more than enough neurons to model the data, there will be a point during training when the training error continues to decrease but the checking error levels off and begins to increase. The script cvtrain is used to check for this behavior. An additional 11 pattern noisy data set will be used as the checking data.

```
x=[0:1:10];
y=2+3*x-.4*x.^2;
randn('seed',5);                                        % Change seed.
tc=2+3*x-.4*x.^2+0.2*randn(1,11).*(2+3*x-.4*x.^2);      % Checking data set.
load data12                                             % Training data set.
save data14 x t tc
plot(x,y,x,t,x,tc);
```

```
title('Training Data and Checking Data')
```

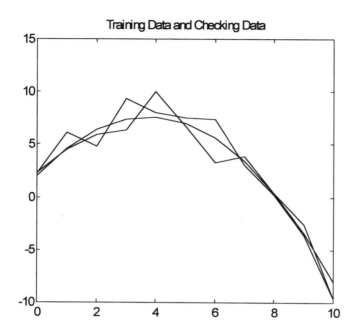

A 5 hidden neuron network will now be trained using cvtrain for 1000 epochs.

*** BP Training complete, error goal not met!
*** Minimum RMS = 6.909391e-001
*** Minimum checking RMS = 1.249887e+000
*** Best Weight Matrix at 240 epochs!

```
load weight16;
clg
semilogy(RMS);
hold on
semilogy(RMSc);
hold off
title('Cross Validation Training Results');
ylabel('Root Mean Squared Error')
text(600,2, 'Checking Error');
text(600,.5,'Training Error');
```

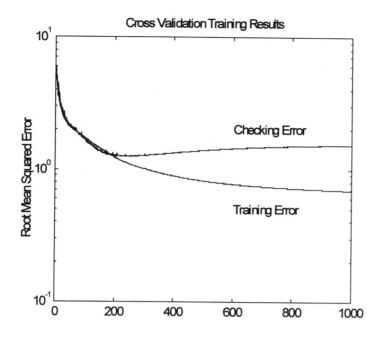

The figure shows that both errors initially are reduced, then the checking error starts to increase. The best weight matrices occur at the checking error minimum. Subsequent to the checking error minimum, the network is overfitting. This training method can also be used to identify a realistic training RMS error goal. In this example, the error goal is the minimum of the checking error.

```
min(RMSc)
```

```
ans =
    1.2499
```

Again we find that a realistic error goal is near 1.2. This agrees with our two previous calculations.

In summary, there are four methods to reduce the chance of overfitting:

1. Limiting the number of free parameters.
2. Training to a realistic error goal.
3. Increase the training set size.
4. Use cross validation training to identify when overfitting occurs.

These methods can be used independently or used together to reduce the chance of overfitting.

Chapter 12 Neural Methods in Fuzzy Systems

12.1 Introduction

In this chapter we explore the use of fuzzy neurons in neural systems while the next chapter we will explore the use of neural methodologies to train fuzzy systems. There is a tremendous advantage to training fuzzy networks when experiential data is available but the advantages of using fuzzy neurons in neural networks is less well defined.

Embedding fuzzy notions into neural networks is an area of active research, and there are few well known or proven results. This chapter will present the fundamentals of constructing neural networks with fuzzy neurons but will not describe the advantages or practical considerations in detail.

12.2 From Crisp to Fuzzy Neurons

The artificial neuron was first presented in Section 7.1. This neuron processed information using the following equation,

$$y = f\left[\sum_{k=1}^{n} x_k w_k + b\right]$$

where x is the input vector, w are the weights connecting the inputs to the neuron and b is a bias. The function is usually continuous and monatonically increasing. The result of the product,

$$d_k = x_k w_k$$

is referred to as the dendritic input to the neuron. In the case of fuzzy neurons, this dendritic input is usually a function of the input and the weight. This function can be an S-norm, such as the probabilistic sum, or a T-norm, such as the product. The choice of the function is dependent on the type of fuzzy neuron being used. Fuzzy neuron types will be discussed in subsequent sections.

S-norms: 1. probabilistic sum: $d_k = x_k \, S \, w_k = x_k + w_k - x_k w_k$

             2. OR: $d_k = x_k \, S \, w_k = x_k \vee w_k = \max(x_k \vee w_k)$

T-norms: 1. product: $d_k = x_k \, T \, w_k = x_k * w_k$

             2. AND: $d_k = x_k \, T \, w_k = x_k \wedge w_k = \min(x_k, w_k)$

The dendritic inputs are then aggregated by some chosen operator. In the most simple case, this operator is a summation operator, but other operators such as min and max are commonly used.

Summation aggregation operator: $I_j = \sum_{k=1}^{n} d_k$

Min aggregation operator: $I_j = \bigwedge_{k=1}^{n} d_k = \min(d_k)$

Max aggregation operator: $I_j = \bigvee_{k=1}^{n} d_k = \max(d_k)$

The fuzzy neuron output y_j is a function of the of the internal activation I_j and the threshold level of the neuron. This function can be a numerical function or a *T*-norm or *S*-norm.

$$y_j = \Phi(I_j, T_j)$$

The use of different operators in fuzzy neurons gives the designer the flexibility to make the neuron perform various functions. For example, the output of a fuzzy neuron may be a linguistic representation of the input vector such as *Small* or *Rapid*. These linguistic representation can be further processed by subsequent layers of fuzzy neurons to model a specified relationship.

12.3 Generalized Fuzzy Neuron and Networks

This section will further discuss the fuzzy neural network architecture. The generalized fuzzy neuron is shown in the figure below.

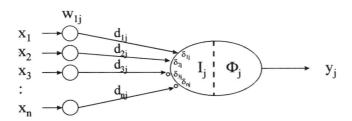

Generalized Fuzzy Neuron

The synaptic inputs (**x**) generally represent the degree of membership to a fuzzy set and have a value in the range [0 1]. The dendritic inputs (**d**) are also normally bounded, in the range [0 1], and represent the membership to a fuzzy set.

In this figure, the dendritic inputs have been modified to produce excitatory and inhibitory signals. This is done with a function that performs the compliment and is represented graphically by the not sign: °. Numerically, the behavior of these operators is represented by:

excitatory $\quad \delta_{ij} = d_{ij}$
inhibitory $\quad \delta_{ij} = 1 - d_{ij}$

The generalized fuzzy neuron has internal signals that represent membership to fuzzy sets. Therefore, these signals have meaning. Most signals in artificial neural networks do not have a discernible meaning. This is one advantage of a fuzzy neural system.

12.4 Aggregation and Transfer Functions in Fuzzy Neurons

As stated in section 12.2, the aggregation operator is implimentable by several different mathematical expressions. The most common aggregation operator is a *T*-norm represented by:

$$I_j = \underset{i=1}{\overset{n}{T}} \delta_{ij}$$

but other operators can also be implemented. The internal activation is simply the aggregation of the modified input membership values (dendritic inputs). A *T*-norm (product or min) tends to reduce the activation while an *S*-norm tends to enhance the activation (probabilistic sum or max). These aggregation operators provide the fuzzy neuron with a means of implementing intersection (*T*-norm) and union (*S*-norm) concepts.

The MATLAB implementation of a *T*-norm aggregation is:

```
x=[0.7  0.3];           % Two inputs.
w=[0.2  0.5];           % Hidden weight matrix.
I=product(x',w')        % Calculate product T-norm output.

I =
    0.1400
    0.1500
```

The MATLAB implementation of a probabilistic sum *S*-norm is:

```
x=[0.7  0.3];           % Two inputs.
w=[0.2  0.5];           % Hidden weight matrix.
I=probor(x',w')         % Calculate probabilistic sum S-norm output.

I =
    0.7600
    0.6500
```

Biases can be implemented in *T*-norm and *S*-norm operations by setting a weight value to either 1 or 0. In normal artificial neurons, the bias is the weight corresponding to a dummy node of input equal to 1. Similarly, in fuzzy neurons, the bias is the weight corresponding to a dummy input value (x_0). For example, a bias in a *T*-norm would have its dummy input set to a 1 while a bias in an *S*-norm would have its dummy input set to a 0.

The MATLAB implementation of a *T*-norm bias is:

```
x=[1      0.3];       % The dummy input is a 1
w=[0.7    0.5];       % The corresponding bias weight is a .7.
I=x'.*w'              % Calculate product T-norm output.

I =
    0.7000
    0.1500
```

The MATLAB implementation of an *S*-norm bias is:

```
x=[0      0.3];       % The dummy input is a 0.
w=[0.7    0.5];       % The corresponding bias weight is a .7.
I=probor(x',w')       % Calculate probabilistic sum S-norm output.

I =
    0.7000
    0.6500
```

In both of the above cases, the bias propagates to the internal activation and may affect the neuron output depending on the activation function operator.

The activation function or transfer function is a mapping operator from the internal activation to the neuron's output. This mapping may correspond to a linguistic modifier. For example if the inputs are weighted grades of accomplishments, the linguistic modifier "*more-or-less*" could give the aggregated value a stronger value. *S*-norms and *T*-norms are commonly used. The MATLAB implementation of an *S*-norm aggregation and a *T*-norm activation function would be:

```
x=[0      0.3];       % The dummy input is a 0.
w=[0.7    0.5];       % The corresponding bias weight is a .7.
I=probor(x',w')       % Calculate probabilistic sum S-norm output.
z=prod(I)             % Calculate product T-norm.

I =
    0.7000
    0.6500
z =
    0.4550
```

12.5 AND and OR Fuzzy Neurons

The most commonly used fuzzy neurons are *AND* and *OR* neurons. The *AND* neuron performs an *S*-norm operation on the dendritic inputs and weights and then performs a *T*-norm operation on the results of the *S*-norm operation. Although any *S*-norm or *T*-norm operations can be implemented, usually max and min operators are used. The exception is when training routines are implemented; in this case, differentiable functions are used. The *AND* neuron representation is:

$$z_h = T_{i=1}^n\left(x_i\, S\, w_{hi}\right)$$

The MATLAB *AND* fuzzy neuron implementation is:

```
x=[0.7  0.3];          % Two inputs.
w=[0.2  0.5];          % Hidden weight matrix.
I=max(x,w)             % Calculate max S-norm operation.
z=min(I)               % Calculate min T-norm operation.

I =
    0.7000    0.5000
z =
    0.5000
```

The *OR* neuron performs a *T*-norm operation on the dendritic inputs and weights and then performs an *S*-norm operation on the results of the *S*-norm operation.

$$z_h = S_{i=1}^n\left(x_i\, T\, w_{hi}\right)$$

The MATLAB *OR* fuzzy neuron implementation is:

```
x=[0.7  0.3];          % Two inputs.
w=[0.2  0.5];          % Hidden weight matrix.
I=min(x,w)             % Calculate min T-norm operation.
z=max(I)               % Calculate max S-norm operation.

I =
    0.2000    0.3000
z =
    0.3000
```

AND and *OR* neurons can be arranged in layers and these layers can be arranged in networks to form multilayer fuzzy neural networks.

12.6 Multilayer Fuzzy Neural Networks

The fuzzy neurons discussed in earlier sections of this chapter can be connected to form multiple layers. The multilayer networks discussed here have three layers with each layer performing a different function. The input layer simply sends the inputs to each of the hidden nodes. The hidden layer is composed of either *AND* or *OR* neurons. These hidden layer neurons perform a norm operation on the inputs and weight matrix. The outputs of the hidden layer neurons perform a norm operation on the hidden layer outputs (z) and the output weight vector (v). The norm operations can be any type of *S*-norms or *T*-norms. The following figure presents a diagram of a multilayer fuzzy network.

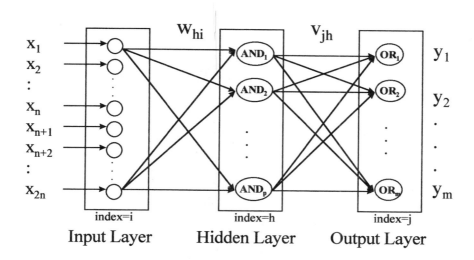

Multilayer Fuzzy Neural Network

In the above figure, the hidden layer uses *AND* neurons to perform the *T*-norm aggregation. The output of the hidden layer is denoted z_h where h is the index of the hidden node.

$$z_h = \left[T_{i=1}^n (x_i \, S \, w_{hi}) \right] T \left[T_{i=1}^n (\bar{x}_i \, S \, w_{h(n+i)}) \right] \quad h = 1, 2, \ldots, p$$

This is implemented in MATLAB with product *T*-norms and probabilistic sum *S*-norms. As a very simple example, suppose we have a network with two inputs, two hidden nodes and one output. The hidden layer outputs are given by:

```
x=[0.1 0.3];              % Two inputs.
xc=[0.9 0.7];             % Two complements.
w= [0.2  0.5; -0.2 0.8];  % Hidden weight matrix.
wc=[0.9 -0.3;  0.6 0.2];  % Complementary portion weight matrix.
for i=1:2                 % Calculate hidden node outputs.
   z(i)=prod([prod(probor(x',w(:,i))) prod(probor(xc',wc(:,i)))]);
end
z

z =
    0.0390    0.3127
```

In the above figure, the output layer uses *OR* neurons to perform an *S*-norm aggregation. The output of the network is denoted y_j where j is the index of the output neuron. In the examples of the text and this supplement, the networks are limited to one output neuron although the MATLAB code provides for multiple outputs.

$$y_j = \left[S_{h=1}^p (z_h \, T \, v_{hj}) \right] \quad j = 1, 2, \ldots, m$$

This is implemented in MATLAB with product *T*-norms and probabilistic sum *S*-norms:

```
v=[0.3 0.7]';          % One output node.
for j=1:2              % Calculate output node internal activations.
  I(j)=prod([z(j) v(j)]');
end
y=probor(I)            % Perform a probor on the internal activations.

y =
    0.2281
```

The evaluation of all the fuzzy neurons in a multilayer fuzzy network can be combined into one function. The following code simulates the fuzzy neural network where:

x is the input vector.	size(x) = (patterns, inputs)
w is the hidden weight vector.	size(w) = (inputs, hidden neurons)
v is the output weight vector.	size(v) = (hidden neurons, output neurons)
y is the output vector.	size(y) = (patterns, output neurons)

As a very simple example, suppose we have a network with two inputs, one hidden node and one output. The forward pass of two patterns gives:

```
x=[.1 .3 .9 .7;.7 .4 .3 .6 ];   % Two patterns of 2 inputs and their
                                % complements.
w=[.2 .5;-.2 .8;.9 -.3;.6 .2;]; % Two hidden nodes.
v=[0.3  0.7]';                  % One output node.
y=fuzzy_nn(x,w,v)               % Simulate the network.

y =
    0.2281
    0.0803
```

12.7 Learning and Adaptation in Fuzzy Neural Networks

If there is experiential input/output data from the relationship to be modeled, the fuzzy neural network's weights and biases can be trained to better model the relationship. This training can be performed with a gradient descent algorithm similar to the one used for the standard neural network.

Complications with fuzzy neural network training arise because the activation or transfer functions must be continuous, monotonically increasing, differentiable operators. Several *S*-norm and *T*-norm operators such as the max and min operators do not fit this requirement, although the probabilistic sum *S*-norm and product *T*-norm that were used in the previous section do meet his requirement.

As a simple example, let us study the weight changes involved in training a single *OR* fuzzy neuron with two inputs. We will assume the *OR* fuzzy neuron is an output neuron and is therefore represented by:

$$y_j = \left[S_{h=1}^{p} \left(z_h \, T \, v_{hj} \right) \right] \quad j=1$$

where: y_j is the jth output neuron.
z_h is the output of the hth hidden layer neuron.
v_{jh} is the weight connecting the hth hidden layer neuron to the jth output.
T-norm is product.
S-norm is probabilistic *OR*

The training data consists of three patterns.

```
z=[1 0.2 0.7;1 0.3 0.9;1 0.1 0.2];  % Training input data with bias.
t=[0.7       ;0.8       ;0.4       ];  % Training output target data.
```

And the output is represented by:

```
v=[.32 .76 .11]              % Random initial weights.
y=zeros(3,1);                % Output vector
for k=1:3                    % For each input pattern
   I=product(z(k,:)',v');    % Calculate product T-norm operation.
   y(k)=probor(I);           % Calculate probabilistic sum S-norm.
end
y


v =
    0.3200    0.7600    0.1100
y =
    0.4678
    0.5270
    0.3855
```

To train this network to minimize a squared error between the target vector (t) and the output vector (y), we will use a gradient descent procedure. The squared error (as opposed to the sum of squared errors) is used because the training will occur sequentially rather in batch mode. The error and squared error are defined as:

$$\varepsilon = t - y$$
$$\varepsilon^2 = (t - y)^2$$

$$\Delta v_{jh} = -\eta \cdot \frac{\partial \varepsilon^2}{\partial v_{jh}}$$

$$= -\eta \cdot \frac{\partial \varepsilon^2}{\partial y_j} \cdot \frac{\partial y_j}{\partial v_{jh}}$$

where:

$$\frac{\partial \varepsilon^2}{\partial y_j} = (-2)[t_j - y_j]$$

$$\frac{\partial y_j}{\partial v_{jh}} = \frac{\partial S_{h=1}^p (z_h T_{jh})}{\partial I_j}$$

and T is the *T*-norm operator (product).
S is the *S*-norm operator (probabilistic sum).

If we substitute the symbol A for the norms of the terms not involving a weight h=q, then we can solve for the partial derivative of y with respect to that weight v_q.

$$A = S_{h \neq q}^p z_h T v_{jh}$$

$$\frac{\partial y_j}{\partial v_{jq}} = \frac{\partial [A + v_{jq} z_q - A v_{jq} z_q]}{\partial v_{jq}} = z_q (1 - A)$$

If the *T*-norm is a product and the *S*-norm is a probabilistic sum, this equation can be written as:

$$\frac{\partial y_j}{\partial v_{jq}} = z_q (1 - A) = z_q \left(1 - \underset{h \neq q}{probor} \left[v_{jh} z_h\right]^p \right)$$

Combining the factor of 2 into the learning rate results in:

$$\Delta v_{jq} = \eta (t_j - y_j) z_q \left(1 - \underset{h \neq q}{probor}\left[v_{jh} z_h\right]^p\right)$$

The error terms and squared error are:

```
error=(t-y)
SSE=sum((t-y).^2)
```

error =

```
        0.2322
        0.2730
        0.0145
SSE =
    0.1287
```

Now lets train the network with a learning rate of 1.

```
lr=1;                                   % Learning rate.
patterns=length(error);                 % Number of input patterns.
cycles=30;                              % Number of training iterations.

% Train the network.
inds=[1:length(v)];
for k=1:cycles
   for j=1:patterns                     % Loop through each pattern.
      for i=1:length(v);                % Loop through each weight.
         ind=find(inds~=i);             % Specify weights other than i.
         A=probor(product(z(j,ind)',v(ind)'));  % Calculate T and S norms.
         delv(i)=lr*error(j)*z(j,i)*(1-A); % Calculate weight update.
      end
      v=v+delv;                         % Update weights.
   end

% Now check the new SSE.
   for k=1:length(error);       % For each input pattern
      I=product(z(k,:)',v');    % Calculate product T-norm operation.
      y(k)=probor(I);           % Calculate probabilitsic sum S-norm.
   end
   error=(t-y);                 % Calculate the eror terms.
   SSE=sum((t-y).^2);           % Find the sum of squared errors.
end
SSE
y

SSE =
  1.3123e-004
y =
    0.6909
    0.8063
    0.4030
```

This closely matches the target vector: t=[.7 .8 .4]. This gradient descent training algorithm can be expanded to train multilayer fuzzy neural networks. The previous section derived the weight update for an fuzzy *OR* output neuron. We will now derive the weight update for a fuzzy multilayer neural network with *AND* hidden units and *OR* output units.

We use the chain rule to find the gradient vector.

$$\frac{\partial \varepsilon^2}{\partial w_{jk}} = \frac{\partial \varepsilon^2}{\partial y_j} \sum_{h=1}^{p} \frac{\partial y_j}{\partial z_h} \frac{\partial z_h}{\partial w_{hi}}$$

and

$$\frac{\partial \varepsilon^2}{\partial y_j} = (-2)[t_j - y_j]$$

Since the inputs and weights in a *AND* and *OR* are treated the same, the partial derivative with respect to the inputs (z_h) is of the same form of the partial derivative with respect to the weights (v_h) that was derived above. Therefore, substituting A for the norms not containing z_q, we get a solution of the same form.

$$A = S_{h \neq q}^{p} z_h T v_{jh}$$

$$\frac{\partial y_j}{\partial z_q} = \frac{\partial [A + v_{jq} z_q - A v_{jq} z_q]}{\partial z_q} = v_{jq}(1 - A)$$

For a product *T*-norm and a probabilistic sum *S*-norm, this results in:

$$\frac{\partial y_j}{\partial z_q} = v_{jq}(1 - A) = v_{jq}\left(1 - probor\left[\underset{h \neq q}{\overset{p}{v_{jh} z_h}}\right]\right)$$

For the second term in the chain rule:

$$\frac{\partial z_h}{\partial w_{hi}} = \frac{\partial \prod_{i=1}^{n}(x_i + w_{hi} - x_i w_{hi})}{\partial w_{hi}}$$

For a specific input weight: w_{hr},

$$\frac{\partial z_h}{\partial w_{hr}} = \frac{\partial}{\partial w_{hr}}\left[\prod_{i \neq r}^{n}(x_i + w_{hi} - x_i w_{hi}) * (x_r + w_{hr} - x_r w_{hr})\right]$$

$$= \frac{\partial}{\partial w_{hi}}\left[\prod_{i \neq r}^{n}(x_i + w_{hi} - x_i w_{hi})x_r + \prod_{i \neq r}^{n}(x_i + w_{hi} - x_i w_{hi})w_{hr} - \prod_{i \neq r}^{n}(x_i + w_{hi} - x_i w_{hi})x_r w_{hr}\right]$$

$$= \prod_{i \neq r}^{n}(x_i + w_{hi} - x_i w_{hi})(1 - x_r)$$

Combining the above chain rule terms results in:

$$\frac{\partial \varepsilon^2}{\partial w_{jk}} = \frac{\partial \varepsilon^2}{\partial y_j} \frac{\partial y_j}{\partial z_h} \frac{\partial z_h}{\partial w_{hi}}$$

$$= \eta \varepsilon \left[\prod_{i \neq r}^{n} (x_i + w_{hi} - x_i w_{hi})(1 - x_r) \right] v_{jq} (1 - probor_{h \neq q}(v_{jh} z_h))$$

As an example of training a multilayer fuzzy neural network, we will consider a network with two hidden *AND* neurons and one output *OR* neuron.

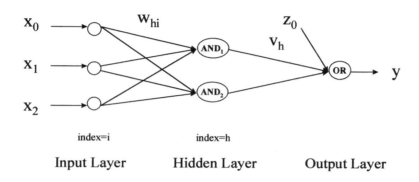

Example Fuzzy Neural Network

In this case we will implement biases in the hidden layer. Since these are *AND* fuzzy neurons, the dummy input must equal 0. A bias will also be implemented in the output node with a dummy input equal to 1. The forward pass results in:

```
clear all;rand('seed',10)
x=[0 0.2 0.7;0 0.3 0.9;0 0.7 0.2];  % Training input data with bias.
t=[0.7      ;0.8      ;0.4      ];  % Training output target data.
w=rand(2,3);               % Random initial hidden layer weights.
v=rand(1,3);               % Random initial output layer weights.
z=zeros(3,2);              % Hidden vector (three patterns, 2 outputs).
y=zeros(3,1);              % Output vector.(three patterns)
for k=1:3                  % For each pattern.
   for h=1:2               % For each hidden neuron.
      z(k,h)=prod(probor(x(k,:)',w(h,:)'));% Output of each hidden node.
   end
   I=product([1 z(k,:)]',v');% Calculate product T-norm operation.
   y(k)=probor(I);         % Calculate probabilitsic sum S-norm.
end
error=(t-y)
SSE=sum((t-y).^2)

error =
    0.2654
    0.2987
```

```
            0.1153
SSE =
            0.1730
```

Now lets train the network with a learning rate of 0.5.

```
lr=0.5;                          % Learning rate.
patterns=size(error,1);          % Number of input patterns.
cycles=20;                       % Number of training iterations.
[h_nodes,inputs]=size(w);        % Define number of inputs and hidden nodes.
indv=[1:size(v,2)];indw=[1:size(w,2)];      % weight matrix indices.

% Update the network output layer.
for m=1:cycles
   for j=1:patterns                          % Loop through patterns.
     for i=1:size(v,2);                      % Loop through weights.
        ind=find(indv~=i);                   % Specify weights not = i.
        Z=[ones(patterns,1) z];              % Output biases.
        A=probor(product(Z(j,ind)',v(ind)'));% Calculate T and S norms.
        delv(i)=lr*error(j)*Z(j,i)*(1-A);    % Calculate weight update.
     end
     v=v+delv;                               % Update weights.
   end

% Update the network hidden layer.
   for j=1:patterns                          % Loop through patterns.
     for l=1:h_nodes
       for i=1:inputs;                       % Loop through weights.
         ind=find(indw~=i);                  % Specify weights not = i.
         A=probor(product(Z(j,l)',v(l)'));   % Calculate OR norms.
         B=prod(probor(x(j,ind)',w(l,ind)'));% Calculate AND norms.
         delw(l,i)=lr*error(j)*B*(1-x(j,i))*v(l)*A;  % Compute update.
       end
     end
     w=w+delw;                               % Update weights.
   end

% Now check the new error terms.
   for k=1:patterns                          % For each pattern.
     for h=1:h_nodes                         % For each input neuron.
        z(k,h)=prod(probor(x(k,:)',w(h,:)'));%Output of hidden nodes.
     end
     I=product([1 z(k,:)]',v'); % Calculate product T-norm operation.
     y(k)=probor(I);                         % Calculate probabilitsic sum S-norm.
   end
   error=(t-y);
end
SSE=sum((t-y).^2)
y

SSE =
    0.0049
y =
    0.6704
```

```
0.7615
0.4502
```

The error has been drastically reduced and the output vector has moved towards the target vector: t=[.7 .8 .4]. By varying the learning rate and training for more iterations, this error can be reduced even further.

Chapter 13 Neural Methods in Fuzzy Systems

13.1 Introduction

The use of neural network training techniques allows us the ability to embed empirical information into a fuzzy system. This greatly expands the range of applications in which fuzzy systems can be used. The ability to make use of both expert and empirical information greatly enhances the utility of fuzzy systems.

One of the limitations of fuzzy systems comes from the curse of dimensionality. Expert information about the relationship to be modeled may make it possible to reduce the rule set from all combinations to the few that are important. In this way, expert knowledge may make a problem tractable.

A limitation of using only expert knowledge is the inability to efficiently tune a fuzzy system to give precise outputs for several, possibly contradictory, input combinations. The use of error based training techniques allows fuzzy systems to learn the intricacies inherent in empirical data.

13.2 Fuzzy-Neural Hybrids

Neural methods can be used in constructing fuzzy systems in ways other than training. They can also be used for rule selection, membership function determination and in what we can refer to as hybrid systems. Hybrid systems are systems that employ both neural networks and fuzzy systems. These hybrid systems may make use of a supervisory fuzzy system to select the best output from several neural networks or may use a neural network to intelligently combine the outputs from several fuzzy systems. The combinations and applications are endless.

13.3 Neural Networks for Determining Membership Functions

Membership function determination can be viewed as a data clustering and classification problem. First the data is classified into clusters and then membership values are given to the individual patterns in the clusters. Neural network architectures such as Kohonen Self Organizing Maps are well suited to finding clusters in input data. After cluster centers are identified, the width parameters of the SOM functions, which are usually Gaussian, can be set so that the SOM outputs are the membership values.

One methodology of implementing this type of two stage process is called the Adeli-Hung Algorithm. The first stage is called classification while the second stage is called fuzzification. Suppose you have N input patterns with M components or inputs. The

Adeli-Hung Algorithm constructs a two layer neural network with M inputs and C clusters. Clusters are added as new inputs, which do not closely resemble old clusters, are presented to the network. This ability to add clusters and allow the network to grow resembles the plasticity inherent in ART networks. The algorithm is implemented in the following steps:

1. Calculate the degree of difference between the input vector X_i and each cluster center C_i. A Euclidean distance can be used:

$$dist(X, C_i) = \sqrt{\sum_{j=1}^{M}(x_j - c_{ij})^2}$$

where: x_j is the jth input.
c_{ij} is the jth component of the ith cluster.
M is the number of clusters.

2. Find the closest cluster to the input pattern and call it C_p.

3. Compare the distance to this closest cluster: $dist(X, C_p)$ with some predetermined distance. If it is closer than the predetermined distance, add it to that cluster, if it is further than the predefined cluster, then add a new cluster and center it on the input vector. When an input is added to a cluster, the cluster center (prototype vector) is recalculated as the mean of all patterns in the cluster.

$$C_p = [c_{p1}\ c_{p2}\ ...c_{pM}] = \frac{1}{n_p}\sum_{i=1}^{n_p} X_i^p$$

4. The membership of an input vector X_i to a cluster C_p is defined as

$$\mu_p = \begin{cases} 0 & \text{if } D^w(X_i^P, C_p) > \kappa \\ 1 - \frac{D^w(X_i^P, C_p)}{\kappa} & \text{if } D^w(X_i^P, C_p) < \kappa \end{cases}$$

where:
κ is the width of the triangular membership function.

$$D^w(X_i^P, C_p) = \sqrt{\sum_{j=1}^{M}(x_{ij}^P - c_{pj})^2}\text{, the weighted norm, is the Euclidean distance:}$$

This results in triangular membership functions with unity membership at the center and linearly decreasing to 0 at the tolerance distance. An example of a MATLAB implementation of the Adeli-Hung algorithm is given below.

```
X=[1 0 .9;0 1 0;1 1 1;0 0 0;0 1 1;.9 1 0;1 .9 1;0 0 .1];% Inputs
C=[1 0 1];         % Matrix of prototype clusters.
data=[1 0 1 1];    % Matrix of data; 4th column specifies class.
tolerance=1;       % Tolerance value used to create new clusters.
for k=1:length(X);
%  Step 1:  Find the Euclidean distance to each cluster center.
   [p,inputs]=size(C);
   for i=1:p
      distance(i)=dist(X(k,:),C(i,:)).^.5;
   end
%  Step 2:  Find the closest cluster.
   ind=find(min(distance)==distance);
%  Step 3:  Compare with a predefined tolerance.
   if min(distance)>tolerance
      C=[C;X(k,:)];                           % Make X a new cluster center.
      data=[data;[X(k,:) p+1]];
   else                                       % Calculate old cluster center.
      data=[data;[X(k,:) ind]];               % Add new data pattern.
      cluster_inds=find(data(:,4)==ind);      % Other clusters in class.
      for j=1:inputs
         C(ind,j)=sum(data(cluster_inds,j))/length(cluster_inds);
      end
   end
%  Step 4:  Calculate memberships to all p classes.
   mu=zeros(p,1);
   for i=1:p
      D=dist(X(k,:),C(i,:)).^.5;
      if D<=tolerance;
         mu(i)=1-D/tolerance;
      end
   end
end
C     % Display the cluster prototypes.
data  % Display all input vectors and their classification.
mu    % Display membership of last input to each prototype.
save AHAdata data C       % Save results for the next section.

C =
    1.0000         0    0.9500
         0    0.3333    0.0333
    0.6667    0.9667    1.0000
    0.9000    1.0000         0
data =
    1.0000         0    1.0000    1.0000
    1.0000         0    0.9000    1.0000
         0    1.0000         0    2.0000
    1.0000    1.0000    1.0000    3.0000
         0         0         0    2.0000
         0    1.0000    1.0000    3.0000
    0.9000    1.0000         0    4.0000
    1.0000    0.9000    1.0000    3.0000
         0         0    0.1000    2.0000
mu =
         0
    0.6601
         0
```

There were four clusters created from the X data matrix and their prototypes vectors were stored in the matrix C. The first three elements in each row of the data matrix is the original data and the last column contains the identifier for the cluster closest to it. The vector mu contains the membership of the last data vector to each of the clusters. It can be seen that it is closest to vector prototype number 2.

The Adeli-Hung Algorithm does a good job of clustering the data and finding their membership to the clusters. This can be used to preprocess data to be input to a fuzzy system.

13.4 Neural Network Driven Fuzzy Reasoning

Fuzzy systems that have several inputs suffer from the curse of dimensionality. This section will investigate and implement the Takagi-Hayashi (T-H) method for the construction and tuning of fuzzy rules, this is commonly referred to as neural network driven fuzzy reasoning. The T-H method is an automatic procedure for extracting rules and can greatly reduce the number of rules in a high dimensional problem, thus making the problem tractable.

The T-H method performs three major functions:

1. Partitions the decision hyperspace into a number of rules. It performs this with a clustering algorithm.
2. Identifies a rule's antecedent values (left hand side membership function). It performs this with a neural network.
3. Identifies a rule's consequent values (right hand side membership function) by using a neural network with supervised training. This part necessitates the existence of target outputs.

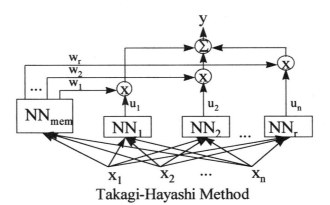

Takagi-Hayashi Method

The above block diagram represents the T-H Method of fuzzy rule extraction. This method uses a variation of the Sugeno fuzzy rule:

if x_i is A_i *AND* x_2 is A_2 *AND* ....x_n is A_n *then* $y=f(x_1, ...x_n)$

where f() is a neural network model rather than a mathematical function. This results in a rule of the form:

if x_i is A_1 AND x_2 is A_2 AND ....x_n is A_n *then* y=NN(x_1, ...x_n).

The NN_{mem} calculates the membership of the input to the *LHS* membership functions and outputs the membership values. The other neural networks form the *RHS* of the rules. The *LHS* membership values weigh the *RHS* neural network outputs through a product function. The altered *RHS* membership values are aggregated to calculate the T-H system output. The neural networks are standard feedforward multilayer perceptron designs.

The following 5 steps implement the T-H Method. The T-H method also implements methods for reducing the neural network inputs to a small set of significant inputs and checking them for overfitting during training.

Step 1: The training data x is clustered into r groups: $R^1, R^2, ..., R^s$ {s=1,2,...,r} with n_t^s terms in each group. Note that the number of inferencing rules will be equal to r.

Step 2: The NN_{mem} neural network is trained with the targets values selected as:

$$w_i^s = \begin{cases} 1, & x_i \in R^s \\ 0, & x_i \notin R^s \end{cases} \quad i=1,...,n_t^s; \quad s=1,...,r$$

The outputs of NN_{mem} for an input x_i are labeled w_i^s, and are the membership values of x_i to each antecedent set R^s.

Step 3: The NN_s networks are trained to identify the consequent part of the rules. The inputs are {$x_{i1}^s,...x_{im}^s$}, and the outputs are y^s i=1,2,...,n_t.

Step 4: The final output value y is calculated with a weighted sum of the NN_s outputs.:

$$y_i = \frac{\sum_{s=1}^{r} \mu_{A^s}(x_i) \cdot \{u_s(x_i)\}_{inf}}{\sum_{s=1}^{r} \mu_{A^s}(x_i)} \quad i=1,2,...n$$

where $u_s(x_i)$ is the calculated output of NN_s.

An example of implementing the T-H method is given below.

Suppose the data and clustering results from the prior example are used. In this example there are 9 input data vectors that have been clustered into four groups:

R^1 has 2 inputs assigned to it ($n_t^1 = 2$).
R^2 has 3 inputs assigned to it ($n_t^2 = 2$).
R^3 has 3 inputs assigned to it ($n_t^3 = 3$).
R^4 has 1 input assigned to it ($n_t^4 = 1$).

Therefore, there are four rules to implement in the system. First we will train the network NN_{mem}.

```
load AHAdata
x=data(:,1:3); % The first three columns of data are the input patterns.
class=data(:,4); % The last column of data is the classification.
t=zeros(9,4);
for i=1:9                % Create the target vector so that the
    t(i,class(i))=1;     % classification column in each target pattern
end                      % is equal to a 1; all other are zero.
t
save TH_data x t

t =
    1    0    0    0
    1    0    0    0
    0    1    0    0
    0    0    1    0
    0    1    0    0
    0    0    1    0
    0    0    0    1
    0    0    1    0
    0    1    0    0
```

We used the bptrain algorithm to train NN_{mem} and save the weights in th_m_wgt. The network's architecture consists of three logistic hidden neurons and a logistic output neuron. Now each of the consequent neural networks (NN_s) must be trained. To train those networks, we must first define the target vectors y and then create the 4 training data sets.

```
% In this example we will define single outputs for each input.
y=[10 12 5 2 4 1 7 1.5 4.5]';
r=4;                     % Number of classifications.
for s=1:r
    ind=find(data(:,4)==s); % Identify the rows in each class.
    x=data(ind,1:3);     % Make the input training matrix.
    t=y(ind,:);          % make the target training matrix.
    eval(['save th_',num2str(s),'_dat x t']);
end
```

The four neural networks with 1 logistic hidden neuron and a linear output neuron were trained off line with bptrain. The weights were saved in th_1_wgt, th_2_wgt, th_3_wgt, and th_4_wgt.

The following code implements the T-H network structure and evaluates the output for a test vector.

```
xt=[-.1 .1 .05]';                                    % Define the test vector.
% Load the neural network weight matrices.
load th_1_wgt;    W11=W1;W21=W2;
load th_2_wgt;    W12=W1;W22=W2;
load th_3_wgt;    W13=W1;W23=W2;
load th_4_wgt;    W14=W1;W24=W2;
load th_m_wgt;    W1n=W1;W2n=W2;
mu = logistic(W2n*[1;logistic(W1n*[1;xt])]);         % NNmem outputs.
% Find the outputs of all four consequent networks.
u1 = W21*[1;logistic(W11*[1;xt])];
u2 = W22*[1;logistic(W12*[1;xt])];
u3 = W23*[1;logistic(W13*[1;xt])];
u4 = W24*[1;logistic(W14*[1;xt])];
u=[u1 u2 u3 u4];        % Vector of NNs outputs.
Ys=u*mu;                % NNs outputs times the membership values (NNmem).
Y=sum(Ys)/sum(mu)       % T-H Fuzzy System ouptut

Y =
   4.4096
```

The closest input to the test vector [-0.1 0.1 0.05] are the vectors [0 1 0], [0 0 0] and [0 0 .1]. These three vectors have target values of 5, 4, and 4.5 respectively, so the output of 4.41 is appropriate. The training vectors were all input to the system and the network performed well for all 9 cases. The T-H method of using neural networks to generate the antecedent and consequent membership functions has been found to be useful and easily implementable with MATLAB.

13.5 Learning and Adaptation in Fuzzy Systems via Neural Networks

When experiential data exists, fuzzy systems can be trained to represent an input-output relationship. By using gradient descent techniques, fuzzy system parameters, such as membership functions (*LHS* or *RHS*), and the connectives between layers in an adaptive network, can be optimized. Adaptation of fuzzy systems using neural network training methods have been proposed by various researchers. Some of the methods described in the literature are: 1) fuzzy system adaptation using gradient-descent error minimization [Hayashi et al. 1992]; 2) optimization of a parameterized fuzzy system with symmetric triangular-shaped input membership functions and crisp outputs using gradient-descent error minimization [Nomura, 1994; Wang, 1994; Jang, 1995]; 3) gradient-descent with exponential MFs [Ichihashi, 1993]; and 4) gradient-descent with symmetric and non-symmetric *LHS* MFs varying connectives and *RHS* forms [Guély, Siarry, 1993].

Regardless of the method or the parameter of the fuzzy system chosen for adaptation, an objective error function, E, must be chosen. Commonly, the squared error is chosen:

$$E = \frac{1}{2}(y - y^t)^2;$$

where y^t is the target output and y is the fuzzy system output. Consider the ith rule of a zero-order Sugeno fuzzy system consisting of n rules (i = 1, ..., n). The figure below presents a zero-order Sugeno system with m inputs and n rules.

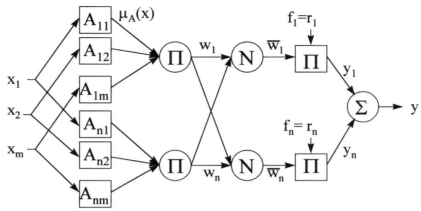

Zero-Order Sugeno System

where: x is the input vector.
A is an antecedent membership function (*LHS*).
$\mu_A(x)$ is the membership of x to set A.
w_i is the degree of fulfillment of the ith rule.
$\overline{w}_i$ is the normalized degree of fulfillment of the ith rule.
r_i is a constant singleton membership function of the ith rule (*RHS*).
y_i is the output of the ith rule.

Mathematically, a zero-order Sugeno system is represented by the following equations:

$$w_i = \prod_{j=1}^{m} \mu_{A_{ji}}(x_j)$$

where $\mu_A(x_j)$ is the membership of x_j to the fuzzy set A, and w_i is the degree of fulfillment of the ith rule. The normalized degree of fulfillment is given by:

$$\overline{w}_i = \frac{w_i}{\sum_{i=1}^{n} w_i}$$

The output (y) of a fuzzy system with *n* rules can be calculated as:

$$y = \sum_{i=1}^{n} \overline{w}_i f_i = \sum_{i=1}^{n} y_i$$

In this case, the system is a zero-order Sugeno system and f_i is defined as:

$$f_i = r_i$$

An example of a 1st order Sugeno system is given in section 13.6.

This notation is slightly different than that in the text, but it is consistent with Jang (1996). When target outputs (y^{rp}) are given, the network can be adapted to reduce an error measure. The adaptable parameters are the input membership functions and the output singleton membership functions, r_i. Now we will look at an example of using gradient descent to optimize the r_i values of this zero-order Sugeno system.

13.5.1 Zero Order Sugeno Fan Speed Control

Consider a zero-order Sugeno fuzzy system used to control the speed of a fan. The objective is to maintain a comfortable temperature in the room based on two input variables: temperature and activity. Three linguistic variables: "*cool*", "*moderate*", and "*hot*" will be used to describe temperature, and two linguistic variables: "*low*" and "*high*" will be used to describe the level of activity in the room. The fan speed will be a crisp output value based on the following set of fuzzy rules:

if Temp is *Cool*	AND Activity is *Low*	*then* Speed is *very_low*	(w_1)
if Temp is *Cool*	AND Activity is *High*	*then* Speed is *low*	(w_2)
if Temp is *Moderate*	AND Activity is *Low*	*then* Speed is *low_medium*	(w_3)
if Temp is *Moderate*	AND Activity is *High*	*then* Speed is *medium*	(w_4)
if Temp is *Hot*	AND Activity is *Low*	*then* Speed is *medium_high*	(w_5)
if Temp is *Hot*	AND Activity is *High*	*then* Speed is *high*	(w_6)

The fuzzy antecedent and consequent membership functions are defined over the universe of discourse with the following MATLAB code.

```
% Universe of Discourse
x = [0:5:100];    % Temperature
y = [0:1:10];     % Activity
z = [0:1:10];     % Fan Speed

% Temperature
cool_mf = mf_trap(x,[0 0 30 50],'n');
moderate_mf = mf_tri(x,[30 55 80],'n');
hot_mf = mf_trap(x,[60 80 100 100],'n');
antecedent_t = [cool_mf;moderate_mf;hot_mf];
plot(x,antecedent_t)
axis([-inf inf 0 1.2]);
title('Antecedent MFs for Temperature')
text(10, 1.1, 'Cool');
text(50, 1.1, 'Moderate');
text(80, 1.1, 'Hot');
xlabel('Temperature')
ylabel('Membership')
```

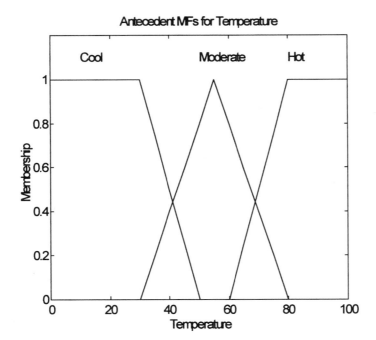

```
low_act = mf_trap(y,[0 0 2 8],'n');
high_act = mf_trap(y,[2 8 10 10],'n');
antecedent_a = [low_act;high_act];
plot(y, antecedent_a);
axis([-inf inf 0 1.2]);
title('Antecedent MFs for Activity')
text(1, 1.1, 'Low');text(8, 1.1, 'High');
xlabel('Activity Level');ylabel('Membership')
```

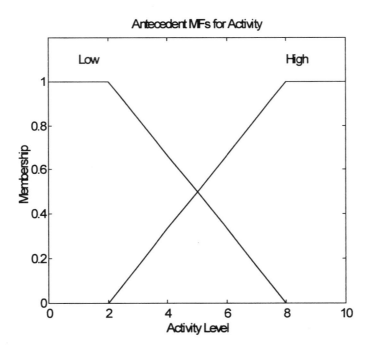

The consequent values of a Sugeno system are crisp singletons. The singletons for the fan speed are defined as:

```
% Fan Speed Consequent Values.
very_low_mf = 1;
low_mf = 2;
low_medium_mf = 4;
medium_mf = 6;
medium_high_mf = 8;
high_mf = 10;
consequent_mf =
[very_low_mf;low_mf;low_medium_mf;medium_mf;medium_high_mf;high_mf];
stem(consequent_mf,ones(size(consequent_mf)))
axis([0 11 0 1.2]);
title('Consequent Values for Fan Speed');
xlabel('Fan Speed')
ylabel('Membership')
text(.5, 1.1, 'Very_Low');
text(2.1, .6, 'Low');
text(3.5, 1.1, 'Low_Medium');
text(6.1, .6, 'Medium');
text(7.5, 1.1, 'Medium_High');
text(10.1, .6, 'High');
```

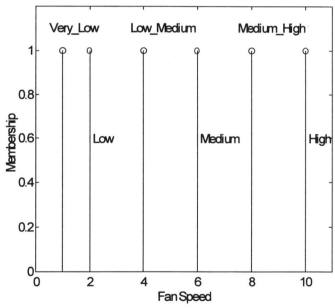

Now that we have defined the membership functions and consequent values, we will evaluate the fuzzy system for an input pattern with temperature = 72.5 and activity = 6.1. First we fuzzify the inputs by finding their membership to each antecedent membership function.

```
temp = 72.5;                                    % Temperature
```

```
mut1=mf_trap(temp, [0 0 30 50],'n');
mut2=mf_tri(temp, [30 55 80],'n');
mut3=mf_trap(temp, [60 80 100 100],'n');
MU_t = [mut1;mut2;mut3]

MU_t =
         0
    0.3000
    0.6250

act = 6.1;                              % Activity
mua1 = mf_trap(act,[0 0 2 8],'n');
mua2 = mf_trap(act,[2 8 10 10],'n');
MU_a = [mua1;mua2]

MU_a =
    0.3167
    0.6833
```

Next, we apply the Sugeno fuzzy *AND* implication operation to find the degree of fulfillment of each rule.

```
antecedent_DOF = [MU_t(1)*MU_a(1)
MU_t(1)*MU_a(2)
MU_t(2)*MU_a(1)
MU_t(2)*MU_a(2)
MU_t(3)*MU_a(1)
MU_t(3)*MU_a(2)]

antecedent_DOF =
         0
         0
    0.0950
    0.2050
    0.1979
    0.4271
```

A plot of the firing strengths of each of the rules is:

```
stem(consequent_mf,antecedent_DOF)
axis([0 11 0 .5]);
title('Consequent Firing Strengths')
xlabel('Fan Speed')
ylabel('Firing Strength')
text(0.2, antecedent_DOF(1)+.02, ' Very_Low');
text(1.5, antecedent_DOF(2)+.04, ' Low');
text(3.0, antecedent_DOF(3)+.02, ' Low_Medium');
text(5.1, antecedent_DOF(4)+.02, ' Medium');
text(7.0, antecedent_DOF(5)+.02, ' Medium_High');
text(9.5, antecedent_DOF(6)+.02, ' High');
```

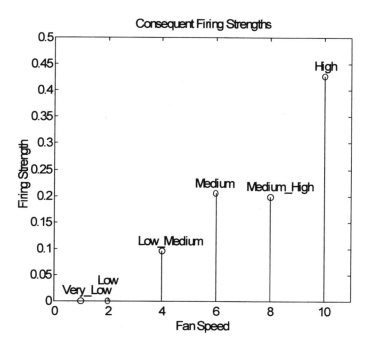

The output is the weighted average of the rule fulfillments.

```
output_y=sum(antecedent_DOF.*consequent_mf)/ sum(antecedent_DOF)
```

```
output_y =
   8.0694
```

The fan speed would be set to a value of 8.07 for a temperature of 72.5 degrees and an activity level of 6.1.

13.5.2 Consequent Membership Function Training

If you desire a specific input-output model and example data are available, the membership functions may be trained to produce the desired model. This is accomplished by using a gradient-descent algorithm to minimize an objective error function such as the one defined earlier.

The learning rule for the output crisp membership functions (r_i) is defined by

$$r_i(t'+1) = r_i(t') - lr \cdot \frac{\partial E}{\partial r_i}$$

where t' is the learning epoch. The update equation can be rewritten as [Nomura et al. 1994]

$$r_i(t'+1) = r_i(t') - lr \cdot \frac{\mu_p^i}{\sum_{i=1}^n \mu_i^p}(y^p - y^{rp}).$$

The m-file *sugfuz.m* demonstrates this use of gradient descent to optimize the output membership functions (r_i).

13.5.3 Antecedent Membership Function Training

The gradient descent algorithm can also be used to optimize the antecedent membership functions. The examples in this supplement use the triangular and trapezoidal membership functions which have non-zero derivatives where the outputs are non-zero. Other adaptable fuzzy systems may use gaussian or generalized bell antecedent membership functions. These functions have non-zero derivatives throughout the universe of discourse and may be easier to implement. Because their derivatives are continuous and smooth, their training performance may also be better. Consider the parameters of a symmetric triangular membership function (a_{ij} = peak; b_{ij} = support):

$$\mu_{ij}(x_j) = \begin{cases} 1 - \dfrac{|x_j - a_{ij}|}{b_{ij}/2}, & if\ |x_j - a_{ij}| \leq \dfrac{b_{ij}}{2} \\ 0, & otherwise \end{cases}$$

A symmetric or non-symmetric triangular membership function can be created using the mf_tri() function, this function will be better described in Section 13.5.4. The membership function can be plotted using the plot function or by using the plot flag. In this example, the plot flag is set to no.

```
x= [0:1:10];                              % universe of discourse
a=5;b=4;                                  % [peak support]
mu_i= mf_tri(x,[a-b/2 a a+b/2],'n');      % calculate memberships
plot(x,mu_i);                             % plot memberships
title('Symmetric Triangular Membership Function');
axis([-inf inf 0 1.2]);xlabel('Input')
ylabel('Membership');text(4.5,1.1,'(a = Peak)');
text(2.5,.1,'(a-0.5b)');text(6.5,.1,'(a+0.5b)');
```

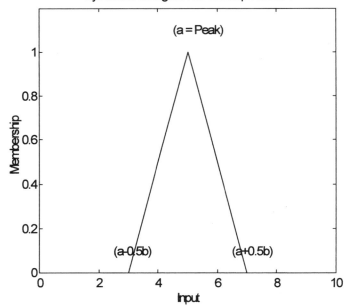

184

The peak parameter update rule is:

$$a_{ij}(t'+1) = a_{ij}(t') - \frac{\eta_a}{p} \cdot \frac{\partial E}{\partial a_{ij}}$$

$$= a_{ij}(t') - \frac{\eta_a}{2p} \cdot \sum_{p'=1}^{p} \frac{\partial E_{p'}}{\partial a_{ij}}$$

where η_a is the learning rate for a_{ij} and p is the number of input patterns. Similar update rules exist for the other MF parameters. The chain rule can be used to calculate the derivatives used to update the MF parameters:

$$\frac{\partial E}{\partial a_{ij}} = \frac{\partial E}{\partial y} \cdot \frac{\partial y}{\partial y_i} \cdot \frac{\partial y_i}{\partial w_i} \cdot \frac{\partial w_i}{\partial \mu_{ij}} \cdot \frac{\partial \mu_{ij}}{\partial a_{ij}}, \quad \text{for the peak parameter;}$$

Using the symmetric triangular MF equations and the product interpretation for *AND*, the partial derivatives are derived below.

$$E = \frac{1}{2}(y - y^t)^2 \quad \text{so} \quad \frac{\partial E}{\partial y} = y - y^t = e$$

$$y = \sum_{i=1}^{n} y_i \quad \text{so} \quad \frac{\partial y}{\partial y_i} = 1$$

$$y_i = \frac{w_i}{\sum_{i=1}^{n} w_i} r_i \quad \text{so} \quad \frac{\partial y_i}{\partial w_i} = \frac{(r_i - y)}{\sum_{i=1}^{n} w_i}$$

$$w_i = \prod_{j=1}^{m} \mu_{A_{ji}}(x_j) \quad \text{so} \quad \frac{\partial w_i}{\partial \mu_{ij}(x_j)} = \frac{w_i}{\mu_{ij}(x_j)}$$

$$\mu_{ij}(x_j) = 1 - \frac{|x_j - a_{ij}|}{b_{ij}/2} \quad \text{so} \quad \frac{\partial \mu_i}{\partial a_{ij}} = \frac{2*sign(x_j - a_{ij})}{b_{ij}} \quad if\ |x_j - a_{ij}| \leq \frac{b_{ij}}{2}$$

$$\text{and} \quad \frac{\partial \mu_i}{\partial a_{ij}} = 0 \quad if\ |x_j - a_{ij}| > \frac{b_{ij}}{2}$$

similarly:

$$\frac{\partial \mu_{ij}}{\partial b_{ij}} = \frac{1 - \mu_{ij}(x_{ij})}{b_{ij}}.$$

Substituting these into the update equation we get

$$\frac{\partial E}{\partial a_{ij}} = (y-y') \cdot \frac{(r_i - y)}{\sum_{i=1}^{n} w_i} \cdot \frac{w_i}{\mu_{ij}(x_j)} \cdot \frac{2*sign(x_j - a_{ij})}{b_{ij}}$$

$$\frac{\partial E}{\partial b_{ij}} = (y-y') \cdot \frac{(r_i - y)}{\sum_{i=1}^{n} w_i} \cdot \frac{w_i}{\mu_{ij}(x_j)} \cdot \frac{1 - \mu_{ij}(x_j)}{b_{ij}}$$

For the RHS membership functions we have:

$$\frac{\partial E}{\partial r_i} = \frac{\partial E}{\partial y} \frac{\partial y}{\partial y_i} \cdot \frac{\partial y_i}{\partial r_i}$$

$$y_i = \overline{w}_i r_i \qquad \text{so} \qquad \frac{\partial y_i}{\partial r_i} = \overline{w}_i$$

resulting in the following gradient.

$$\frac{\partial E}{\partial w_i} = (y - y') \cdot 1 \cdot \overline{w}_i .$$

13.5.4 Membership Function Derivative Functions

Four functions were created to calculate the MF values and derivatives of the MF with respect to their parameters for non-symmetric triangular and trapezoidal MFs. They are *mf_tri.m*, *mf_trap.m*, *dmf_tri.m* and *dmf_trap.m*.

The MATLAB code used in the earlier Fuzzy Logic Chapters required the input to be a defined point on the universe of discourse. These functions do not have limitations on the input and are therefore more general in nature. For example, if the universe of discourse is defined as x=[0:1:10], earlier functions could only be evaluated at those points and an error would occur for an input such as input=2.3. The functions described in this section can handle all ranges of inputs.

Let's look at the triangular membership function outputs and derivatives first.

The function mf_tri can construct symmetrical or non-symmetric triangular membership functions.

```
x=[-10:.2:10];
[mf1]=mf_tri(x,[-4 0 3],'y');
```

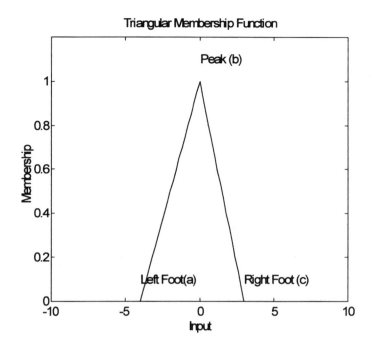

Derivative of non-symmetric triangular MF with respect to its parameters

```
[mf1]=dmf_tri(x,[-4 0 3],'y');
```

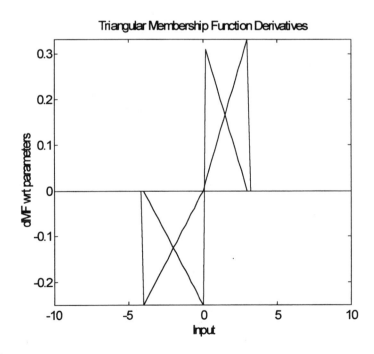

Trapezoidal membership functions:

```
[mf1]=mf_trap(x,[-5 -3 3 5],'y');
```

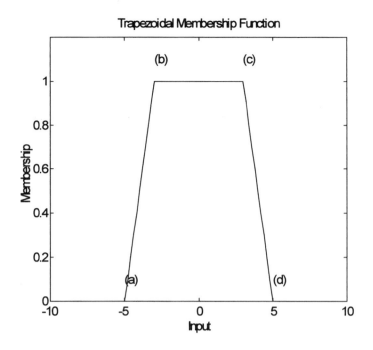

Derivative of trapezoidal MF with respect to it parameters:

```
[mf1]=dmf_trap(x,[-5 -3 3 5],'y');
```

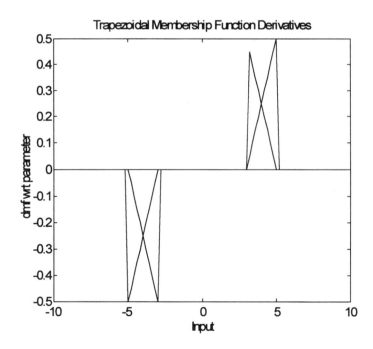

The m-file *adaptri.m* demonstrates the use of gradient descent to optimize parameters of 3 non-symmetric triangular MFs and consequent singletons to approximate the function

$$f(x) = 0.05*x^3 - 0.02*x^2 - 0.3*x + 20.$$

The m-file *adpttrap.m* demonstrates the use of gradient descent to optimize parameters of 3 trapezoid MFs and consequent singletons to approximate the same function.

13.5.5 Membership Function Training Example

As a simple example of using gradient descent to adjust the parameters of the antecedent and singleton consequent membership functions, consider the following problem. Assume that a large naval gun is properly sighted when it is pointing at 5 degrees. The universe of discourse will be defined from 1 to 10 degrees, and the membership to "*Sighted Properly*" will be defined as:

```
clear all;x= [0:.5:10];          % input
num_pts=size(x,2);               % # of points
a=5;                             % peak
b=4;                             % support
mu_i=mf_tri(x,[a-b/2 a a+b/2],'y');
title('MF representing "Sighted Properly"');
xlabel('Direction in Degrees')
ylabel('Membership')
```

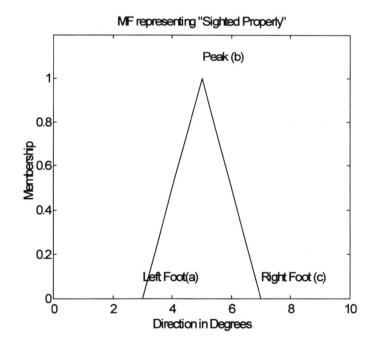

To decide if an artillery shell will hit the target (on a scale of 1-10), consider the following zero-order Sugeno fuzzy system with one rule and r = 10.

if   Gun is *Sighted Properly*   *then*   "Chance of Hit" is 10   (r=10).

Suppose we have experimentally calculated the input-output surface to be:

```
r=10;                            % r value of zero-order Sugeno rule
```

```
y_t=mu_i*r;                              % Chance of Hit
plot(x,y_t);
title('Input-Output Surface for Gun');
axis([-inf inf 0 10.5]);
xlabel('Direction (Input)')
ylabel('Chance of Hit (Output)')
```

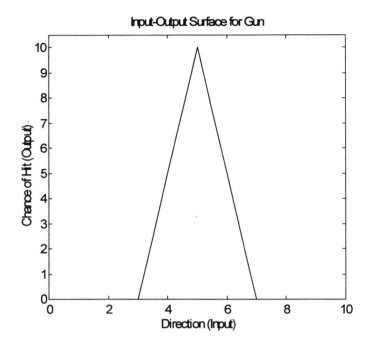

We will now demonstrate how the antecedent MF parameters of a symmetric triangular MF and the consequent r value can be optimized using gradient descent. We will use the input-output data above as training data and initialize the MFs to $a=3$, $b=4$, and $r=8$. The general procedure will involve these steps:

1. A forward pass of all inputs to calculate the output;
2. The calculation of the error and SSE for all inputs;
3. The calculation of the gradient vectors;
4. The updating of the parameters based on the update rule; and
5. Repeating steps 1 through until the training error goal is reached or the maximum number of training epochs is exceeded.

Step 1: Forward pass of all inputs to calculate the output
```
% Initial MF parameters
r=8;                                     % initial r value
a=3;                                     % initial peak value
b=6;                                     % initial support value
mu_i=mf_tri(x,[a-b/2 a a+b/2],'n');      % initial MFs
y=mu_i*r;                                % initial output
plot(x,y_t,x,y);
title('Input-Output Surface');
axis([-inf inf 0 10.5]);
```

```
text(2.5, 4.5, 'Initial')
text(4.5,6, 'Target')
xlabel('Direction (Input)')
ylabel('Chance of Hit (Output)')
```

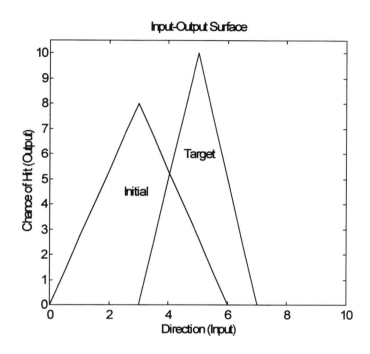

Step 2: Calculate the error and SSE for all inputs
```
e=y-y_t;
SSE=sum(sum(e.^2))

SSE =
   314.5556
```

Step: 3 Calculate the gradient vectors (note: one input, one rule)
```
ind=find(abs(x-a)<=(b/2));    % Locate indices under MF.
delta_a=r*e(ind).*((2*sign(x(ind)-a))/b);% Deltas for ind points.
delta_b=r*e(ind).*((1-mu_i(ind))/b);
delta_r=e(ind).*mu_i(ind);
```

Step 4: Update the parameters with the update rule
```
lr_a=.1;
lr_b=5;
lr_r=10;
del_a=-((lr_a/(2*num_pts))*sum(delta_a));
del_b=-((lr_b/(2*num_pts))*sum(delta_b));
del_r=-((lr_r/(2*num_pts))*sum(delta_r));
a=a+del_a;
b=b+del_b;
r=r+del_r;

a =
    3.7955
```

```
b =
   4.3239
r =
   5.8409
```

Let's see if the SSE is any better:

```
mu_i=mf_tri(x,[a-b/2 a a+b/2],'n');
y_new=mu_i*r;                           % Ready To Eat
e=y_t-y_new;                            % error
SSE(2)=sum(sum(e.^2))                   % Sum Squared error

SSE =
  314.5556   189.1447
```

The SSE was reduced from 314 to 190 in one pass. Lets now iteratively train the fuzzy system.

Step 5: Repeat steps 1 through 5

```
maxcycles=30;
SSE_goal=.5;
for i=2:maxcycles
  mu_i=mf_tri(x,[a-b/2 a a+b/2],'n');
  y_new=mu_i*r;                         % Output
  e=y_new-y_t;                          % Output Error
  SSE(i)=sum(sum(e.^2));                % SSE
  if SSE(i) < SSE_goal; break;end
  ind=find(abs(x-a)<=(b/2));
  delta_a=r*e(ind).*((2*sign(x(ind)-a))/b);
  delta_b=r*e(ind).*((1-mu_i(ind))/b);
  delta_r=e(ind).*mu_i(ind);
  del_a=-((lr_a/(2*num_pts))*sum(delta_a));
  del_b=-((lr_b/(2*num_pts))*sum(delta_b));
  del_r=-((lr_r/(2*num_pts))*sum(delta_r));
  a=a+del_a;
  b=b+del_b;
  r=r+del_r;
end
```

Now lets plot the results:

```
plot(x,y_t,x,y_new,x,y);
title('Input-Output Surface');
axis([0 10 0 10.5]);
text(2.5,5, 'Initial');
text(4.5,7.5, 'Target');
text(4,6.5, 'Final');
xlabel('Input');ylabel('Output');
```

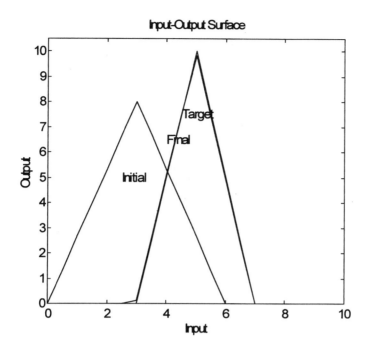

Plot SSE training performance:

```
semilogy(SSE);title('Training Record SSE')
xlabel('Epochs');ylabel('SSE');grid;
```

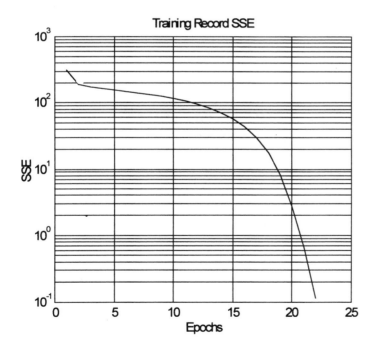

The m-file *adaptfuz.m* demonstrates the use of gradient descent to optimize the MF parameters and shows how the membership functions and input-output relationship changes during the training process.

13.6 Adaptive Network-Based Fuzzy Inference Systems

Jang and Sun [Jang, 1992, Jang and Gulley, 1995] introduced the adaptive network-based fuzzy inference system (ANFIS). This system makes use of a hybrid learning rule to optimize the fuzzy system parameters of a first order Sugeno system. A first order Sugeno system can be graphically represented by:

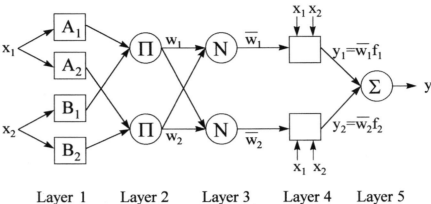

ANFIS architecture for a two-input, two-rule first-order Sugeno model with [Jang 1995a].

where the consequence parameters (p, q, and r) of the nth rule contribute through a first order polynomial of the form:

$$f_n = p_n x_1 + q_n x_2 + r_n$$

13.6.1 ANFIS Hybrid Training Rule

The ANFIS architecture consists of two trainable parameter sets:

1). The antecedent membership function parameters [a,b,c,d].
2). The polynomial parameters [p,q,r], also called the consequent parameters.

The ANFIS training paradigm uses a gradient descent algorithm to optimize the antecedent parameters and a least squares algorithm to solve for the consequent parameters. Because it uses two very different algorithms to reduce the error, the training rule is called a hybrid. The consequent parameters are updated first using a least squares algorithm and the antecedent parameters are then updated by backpropagating the errors that still exist.

The ANFIS architecture consists of five layers with the output of the nodes in each respective layer represented by O_i^l, where i is the ith node of layer l. The following is a layer by layer description of a two input two rule first-order Sugeno system.

Layer 1. Generate the membership grades:

$$O_i^1 = \mu_{A_i}(x)$$

Layer 2. Generate the firing strengths.

$$O_i^2 = w_i = \prod_{j=1}^{m} \mu_{A_i}(x)$$

Layer 3. Normalize the firing strengths.

$$O_i^3 = \overline{w}_i = \frac{w_i}{w_1 + w_2}$$

Layer 4. Calculate rule outputs based on the consequent parameters.

$$O_i^4 = y_i = \overline{w}_i f_i = \overline{w}_i (p_i x_1 + q_i x_2 + r_i)$$

Layer 5. Sum all the inputs from layer 4.

$$O_1^5 = \sum_i y_i = \sum_i \overline{w}_i f_i = (\overline{w}_1 x_1) p_1 + (\overline{w}_1 x_2) q_1 + \overline{w}_1 r_1 + (\overline{w}_2 x_2) p_2 + (\overline{w}_2 x_2) q_2 + \overline{w}_2 r_2$$

It is in this last layer that the consequent parameters can be solved for using a least square algorithm. Let us rearrange this last equation into a more usable form:

$$O_1^5 = y = (w_1 x_1) p_1 + (w_1 x_2) q_1 + w_1 r_1 + (w_2 x_1) p_2 + (w_2 x_2) q_2 + w_2 r_2$$

$$y = \begin{bmatrix} w_1 x_1 & w_1 x_2 & w_3 & w_2 x_1 & w_2 x_2 & w_2 \end{bmatrix} \begin{bmatrix} p_1 \\ q_1 \\ r_1 \\ p_2 \\ q_2 \\ r_2 \end{bmatrix} = \mathbf{XW}$$

When input-output training patterns exist, the weight vector ($\mathbf{W}$), which consist of the consequent parameters, can be solved for using a regression technique.

13.6.2 Least Squares Regression Techniques

Since a least squares technique can be used to solve for the consequent parameters, we will investigate three different techniques used to solve this type of problem. When the output layer of a network or system performs a linear combination of the previous layer's output, the weights can be solved for using a least squares method rather than iteratively trained with a backpropagation algorithm.

The rule outputs are represented by the p by 3n dimensional matrix $\mathbf{X}$, with p rows equal to the number of input patterns and n columns equal to the number of rules. The desired or target outputs are represented by the p by m dimensional matrix **Y** with p rows and m columns equal to the number of outputs. Setting the problem up in this format allows us to use a lease squares solution for the weight vector **W** of n by m dimension..

Y=X*W

If the **X** matrix were invertable, it would be easy to solve for **W**:

W=X^{-1}*Y

but this is not usually the case. Therefore, a pseudoinverse can be used to solve for **W**:

W=(X^T*X)$^{-1}$*X^T*Y   (where **X^T** is the transpose of **X**)

This will minimize the error between the predicted **Y** and the target **Y**. However, this method involves inverting (**X'*X**), which can cause numerical problems when the columns of x are dependent. Consider the following regression example.

```
% Pseudoinverse Solution

x=[2 1
4 2
6 3
8 4];

y=[3 6 9 12]';

w=inv(x'*x)*x'*y;
sse=sum(sum((x*w-y).^2))
```

Warning: Matrix is singular to working precision.

```
sse =
   Inf
```

Although dependent rows can be removed, data is rarely dependent. Consider data that has a small amount of noise that may result in a marginally independent case. We will use the marginally independent data for training and the noise free data to check for generalization.

```
X=[2.000000001  1
3.999999998 2
5.999999998 3
8.000000001 4];

w=inv(X'*X)*X'*y
SSE=sum(sum((X*w-y).^2))
sse=sum(sum((x*w-y).^2))
```

Warning: Matrix is close to singular or badly scaled.

```
                  Results may be inaccurate.  RCOND = 7.894919e-018
w =
    -0.3742
     0.7496
SSE =
   269.7693
sse =
   269.7693
```

The warning tells us that numerical instabilities may result from the nearly singular case and the SSE shows that instabilities did occur. A case with independent patterns results in no errors or warnings.

```
X1=[2.001  1
4.00  2
6.00  3
8.00  4.001];

% Pseudoinverse Solution
w=inv(X1'*X1)*X1'*y
SSE=sum(sum((X1*w-y).^2))
sse=sum(sum((x*w-y).^2))

w =
     0.9504
     1.0990
SSE =
   1.1583e-006
sse =
   9.5293e-007
```

Better regression methods use the LU (Lower Triangular, Upper Triangular) or more robust QR (Orthogonal, Right Triangular) decompositions rather than a simple inversion of the matrix, and the best method uses the singular value decomposition (SVD) [Masters 1993]. The MATLAB command / uses a QR decomposition with pivoting. It provides a least squares solution to an under or over-determined system.

```
% QR Decomposition

w=(y'/X')'
sse=sum(sum((X*w-y).^2))
SSE=sum(sum((x*w-y).^2))

w =
     0.0000
     3.0000
sse =
   7.2970e-030
SSE =
   7.2970e-030
```

The SVD method of solving for the output weights also has the advantage of giving the user control to remove unimportant information that may be related to noise. By removing this unimportant information, one lessens the chance of overfitting the function to be approximated. The SVD method decomposes the x matrix into a diagonal matrix **S** of the same dimension as **X** that contains the singular values, and unitary matrix **U** of principle components, and an orthonormal matrix of right singular values **V**.

$$X = U\ S\ V^T$$

The singular values in **S** are positive and arranged in decreasing order. Their magnitude is related to the information content of the columns of **U** (principle components) that span **X**. Therefore, to remove the noise effects on the solution of the weight matrix, we simply remove the columns of **U** that correspond to small diagonal values in **S**. The weight matrix is then solved for using:

$$W = V\ S^{-1}\ U^T Y$$

The SVD methodology uses the most relevant information to compute the weight matrix and discards unimportant information that may be due to noise. Application of this methodology has sped up the training of neural networks 40 fold [Uhrig et. al. 1996] and resulted in networks with superior generalization capabilities.

```
% Singular Value Decomposition (SVD)

[u,s,v]=svd(X,0);;
inv_s=inv(s)

inv_s =
  1.0e+008 *
    0.0000         0
         0    7.3855
```

We can see that the first singular value has very little information in it. Therefore, we discard its corresponding column in U. This discards the information related to the noise and keeps the solution from attempting to fit the noise.

```
for i=1:2
   if s(i,i)<.1
      inv_s(i,i)=0;
   end
end
w=v*inv_s*u'*y
sse=sum(sum((x*w-y).^2))
SSE=sum(sum((X*w-y).^2))

w =
    1.2000
    0.6000
sse =
```

```
  1.2000e-018
SSE =
  1.3200e-017
```

The SVD method did not reduce the error as much as did the QR decomposition method, but this is because the QR method tried to fit the noise. Remember that this is overfitting and is not desired. MATLAB has a pinv() function that automatically calculates a pseudo-inverse the SVD.

```
w=pinv(X,1e-5)*y
sse=sum(sum((x*w-y).^2))
SSE=sum(sum((X*w-y).^2))

w =
    1.2000
    0.6000
sse =
  1.2000e-018
SSE =
  1.3200e-017
```

We see that the results of the two SVD methods are close to identical. The pinv function will be used in the following example.

13.6.3 ANFIS Hybrid Training Example

To better understand the ANFIS architecture and training paradigm consider the following example of a first order Sugeno system with three rules:

if   x is A_1 *then* $f_1 = p_1 x + r_1$
if   x is A_2 *then* $f_2 = p_2 x + r_2$
if   x is A_3 *then* $f_3 = p_3 x + r_3$

This ANFIS has three trapezoidal membership functions (A_1, A_2, A_3) and can be represented by the following diagram:

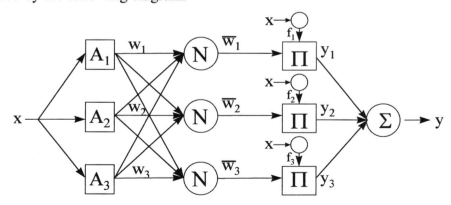

ANFIS architecture for a one-input first-order Sugeno fuzzy model with three rules

First we will create training data for the function to be approximated:

$$f(x) = 0.05*x^3 - 0.02*x^2 - 0.3*x + 20.$$

```
x=[-10:1:10];clg;
y_rp=.05*x'.^3-.02*x'.^2-.3*x'+20;
num_pts=size(x,2);              % number of input patterns
plot(x,y_rp);
xlabel('Input');
ylabel('Output');
title('Function to be Approximated');
```

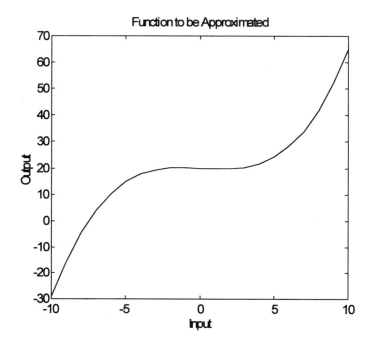

Now we will step through the ANFIS training paradigm.

Layer 1. Generate the membership grades:

$$O_i^1 = \mu_A(x)$$

```
% LAYER 1   MF values
% Initialize Antecedent Parameters
a1=-17; b1=-13; c1=-7; d1=-3;
a2=-9; b2=-4; c2=3; d2=8;
a3=3; b3=7; c3=13; d3=8;
[mf1]=mf_trap(x,[a1 b1 c1 d1],'n');
[mf2]=mf_trap(x,[a2 b2 c2 d2], 'n');
[mf3]=mf_trap(x,[a3 b3 c3 d3], 'n');

% Plot MFs
plot(x,mf1,x,mf2,x,mf3);
```

```
axis([-inf inf 0 1.2]);
title('Initial Input MFs')
xlabel('Input');
ylabel('Membership');
text(-9, .7, 'mf1');
text(-1, .7, 'mf2');
text(8, .7, 'mf3');
```

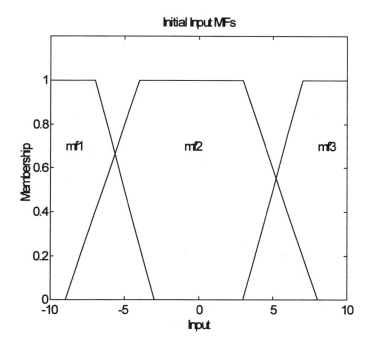

Layer 2. Generate the firing strengths.

$$O_i^2 = w_i = \mu_{A_i}(x)$$

```
% LAYER 2 calculates the firing strength of the
% 1st order Sugeno rules.  Since each rule only has one antecedent
% membership function, no product operation is necessary.
w1=mf1;                  % rule 1
w2=mf2;                  % rule 2
w3=mf3;                  % rule 3
```

Layer 3. Normalizes the firing strengths.

$$O_i^3 = \overline{w}_i = \frac{w_i}{w_1 + w_2 + w_3}$$

```
% LAYER 3
% Determines the normalized firing strengths for the rules (nw) and
% sets to zero if all rules are zero (prevent divide by 0 errors)
for j=1:num_pts
   if (w1(:,j)==0 & w2(:,j)==0 & w3(:,j)==0)
```

```
      nw1(:,j)=0;
      nw2(:,j)=0;
      nw3(:,j)=0;
   else
      nw1(:,j)=w1(:,j)/(w1(:,j)+w2(:,j)+w3(:,j));
      nw2(:,j)=w2(:,j)/(w1(:,j)+w2(:,j)+w3(:,j));
      nw3(:,j)=w3(:,j)/(w1(:,j)+w2(:,j)+w3(:,j));
   end
end
```

Calculate the consequent parameters.

```
X_inner=[nw1.*x;nw2.*x;nw3.*x;nw1;nw2;nw3];
C_parms=pinv(X_inner')*y_rp   % [p1 p2 p3 r1 r2 r3]

C_parms =
    8.3762
    0.7055
    8.2927
   57.4508
   20.0514
  -21.1564
```

Layers 4 and 5. Calculate the outputs using the consequent parameters.

$$O_i^4 = \overline{w}_i f_i = \overline{w}_i (p_i x + r_i)$$
$$O_1^5 = \sum_i \overline{w}_i f_i = (w_1 x)p_1 + w_1 r_1 + (w_2 x)p_2 + w_2 r_2 + (w_3 x)p_3 + w_3 r_3$$

```
% LAYERS 4 and 5
% Calculate the outputs using the inner layer outputs and the
% consequent parameters.
y=X_inner'*C_parms;
```

Plot the Results.

```
plot(x,y_rp,'+',x,y);
xlabel('Input','fontsize',10);
ylabel('Output','fontsize',10);
title('Function Approximation');
legend('Reference','Output')
```

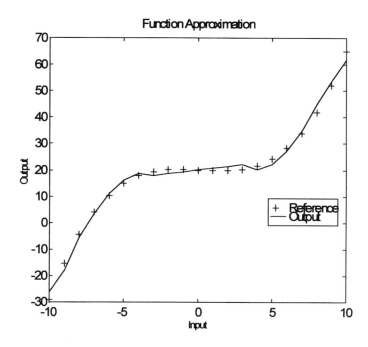

We can see that by just solving for the consequent parameters we have a very good approximation of the function. We can also train the antecedent parameters using gradient descent.

Each ANFIS training epoch, using the hybrid learning rule, consists of two passes. The consequent parameters are obtained during the forward pass using a least-squares optimization algorithm and the premise parameters are updated using a gradient descent algorithm. During the forward pass all node outputs are calculated up to layer 4. At layer 4 the consequent parameters are calculated using a least-squares regression method. Next, the outputs are calculated using the new consequent parameters and the error signals are propagated back through the layers to determine the premise parameter updates.

The consequent parameters are usually solved for at each epoch during the training phase, because as the output of the last hidden layer changes due the backpropagation phase, the consequent parameters are no longer optimal. Since the SVD is computationally intensive, it may be most efficient to perform it every few epochs versus every epoch.

The m-file *anfistrn.m* demonstrates use of the hybrid learning rule to train an ANFIS architecture to approximate the function mentioned above. The consequent parameters (`[p1 p2 p3 r1 r2 r3]`) after the first SVD solution were:

C_parms = [8.8650 4.5311 8.6415 64.2963 23.7368 -26.0706]

and the final consequent parameters were:

C_parms =[11.0308 7.3120 10.5315 82.7411 25.4364 -41.2865]

Below is a graph of the initial antecedent membership functions and the function approximation after one iteration.

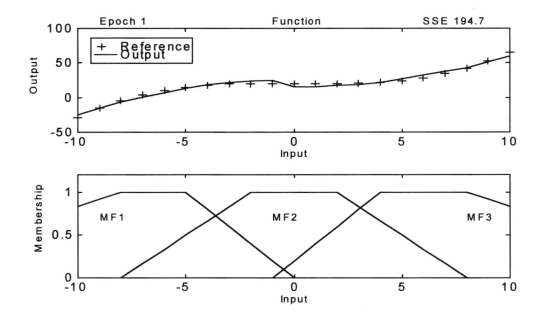

Below is a graph of the final antecedent membership functions and the function approximation after training is completed. The sum of squared error was reduced from 200 to 13 in 40 epochs.

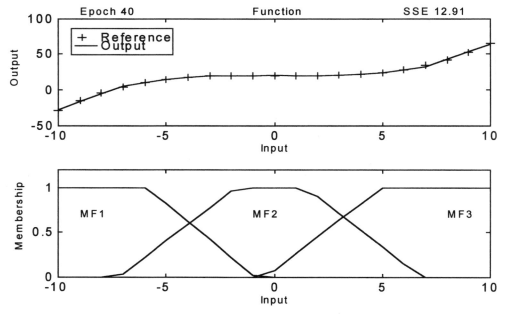

A graph or the training record shows that the ANFIS was able to learn the input output patters with a high degree of accuracy. The *anfistrn.m* code only updates the consequent parameters every 10 epochs in order to reduce the SVD computation time. This results in a training record with dips every 10 epochs. Updating the consequent parameters every epoch did not provide significant error reductions improvements and slowed down the training process.

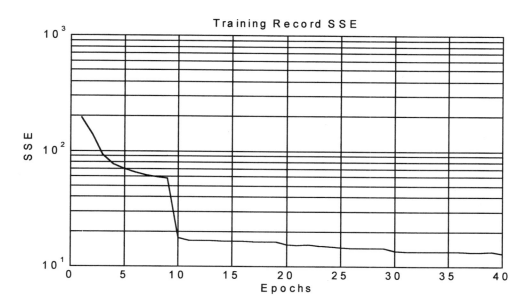

Chapter 14 General Hybrid Neurofuzzy Applications

No specific hybrid neurofuzzy applications will be examined in this supplement, although Chapters 12 and 13 present the methodologies and tools necessary to implement them. The application of the tools and techniques described in Chapter 14 of *Fuzzy and Neural Approaches in Engineering*, is left to the reader.

Chapter 15 Dynamic Hybrid Neurofuzzy Systems

Chapter 15 of *Fuzzy and Neural Approaches in Engineering*, presents several hybrid Neurofuzzy Systems that were developed at The University of Tennessee. These applications are complex and can not be easily implemented with the introductory tools developed in this Supplement. Therefore, for further information on these subjects, the reader should consult the references given in the text.

Chapter 16 Role of Expert Systems in Neurofuzzy Systems

MATLAB does not have an expert system toolbox, nor is a user contributed toolbox available. Since Fuzzy Systems may be viewed as a special type of expert systems that handle uncertainty well, expert systems could be generated by using the fuzzy systems tools with crisp membership functions.

MATLAB does have if->then programming constructs, so expert rules containing heuristic knowledge can be embedded in the neural and fuzzy systems described in earlier chapters; although no examples will be given in this supplement.

Chapter 17 Genetic Algorithms

MATLAB code will not be used to implement Genetic Algorithms in this supplement. Commercial and user Genetic Algorithm toolboxes are available for use with MATLAB. The following is information on a user contributed toolbox and a commercially available toolbox.

GENETIC is a user contributed set of genetic algorithm m-files written by Andrew F. Potvin of The MathWorks, Inc. His email is potvin@mathwork.com. This set of m-files tries to maximize a function using a simple genetic algorithm. It is located at: http://www.mathworks.com/optims.html

FlexTool(GA) is a commercially available package, the following is an excerpt from their email advertising:
FlexTool(GA) M 1.1 Features:
 - Modular, User Friendly, Hardware and operating system transparent
 - Expert, Intermediate, and Novice help settings
 - Hands-on tutorial with step by step application guidelines
 - Designed to draw on MATLAB power
 - Cold Start (start using previously selected GA parameters)
 - Warm Start (start from the previous generation) features
 - GA options : generational GA, steady state GA, micro GA
 - Coding schemes include binary, logarithmic, and real
 - Selection strategies : tournament, roulette wheel, ranking
 - Crossover techniques include 1, 2, multiple point crossover
 - Niching module to identify multiple solutions
 - Clustering module : Use separately or with Niching module
 - Can optimize multiple objectives
 - Default parameter settings for the novice
 - Statistics, figures, and data collection
For more information, contact:
   Flexible Intelligence Group, L.L.C., Box 1477
   Tuscaloosa, AL 35486-1477, USA
   Voice: (205) 345-5166
   Fax : (205) 345-5095
   email: FIGLLC@AOL.COM

References

Demuth, H. and M. Beale, *Neural Networks Toolbox*, The MathWorks Inc., Natick, MA., 1994.

DeSilva, C. W., *Intelligent Control: Fuzzy Logic Applications*, CRC Press, Boca Raton, Fl, 1995.

Elman, J., *Finding Structure in Time*, Cognitive Science, Vol. 14, 1990, pp. 179-211.

Grossberg, S., *Studies of the Mind and Brain*, Reidel Press, Drodrecht, Holland, 1982.

Guely, F., and P. Siarry, Gradient Descent Method for Optimizing Various Fuzzy Rules, *Proceedings of the Second IEEE International Conference on Fuzzy Systems*, San Francisco, 1993, pp. 1241-1246.

Hagan, M., H. Demuth, and Mark Beale, *Neural Network Design*, PWS Publishing Company, Boston, MA, 1996.

Hebb, D., *Organization of Behavior*, John Wiley, New York, 1949.

Hertz, J., A. Krough, R. G. Palmer, *Introduction to the Theory of Neural Computing*, Redwood City, California, Addison-Wesley, 1991.

Hanselman, D. and B. Littlefield, *Mastering MATLAB*, Prentice Hall, Upper Saddle River, NJ, 1996.

Hayashi, I., H. Nomura, H. Yamasaki, and N., Wakami, Construction of Fuzzy Inference Rules by NDF and NDFL, *International Journal of Approximate Reasoning*, Vol. 6, 1992, pp. 241-266.

Ichihashi, H., T. Miyoshi, and K. Nagasaka, Computed Tomography by Neuro-fuzzy Inversion, in *Proceedings of the 1993 International Joint Conference on Neural Networks*, Part 1, Nagoya, Oct 25-29, 1993, pp. 709-712.

Irwin, G. W., K. Warwock, and K. J. Hunt, *Neural Network Applications in Control*, Institute of Electrical Engineers, London, 1995.

Jamshidi, M., N. Vadiee, and T. J. Ross (Eds.), *Fuzzy Logic and Control*, edited by, PTR Prentice Hall, Englewood Cliffs, NJ, 1993.

Jang, J.-S., and N. Gulley, *Fuzzy Logic Toolbox for Use with MATLAB*, The MathWorks Inc., Natick, MA, 1995.

Jang, J.-S., C.-T. Sun, Neuro-fuzzy Modeling and Control, *Proceedings of the IEEE*, Vol. 83, No. 3, March, 1995, pp. 378-406.

Jang, J.-S., C.-T. Sun, and E. Mizutani, *Neuro-Fuzzy and Soft Computing*, Prentice Hall, Upper Saddle River, NJ, 1997.

Jordan, M., Attractor Dynamics and Parallelism in a Connectionist Sequential Machine, *Proc. of the Eighth Annual Conference on Cognitive Science Society*, Amherst, 1986, pp. 531-546.

Kandel, A., and G. Langholz, *Fuzzy Control Systems*, CRC Press, Boca Ratton, Fl, 1994.

Kohonen, T., *Self-Organization and Associative Memory*, Springer-Verlag, Berlin, 1984.

Kosko, B., Bidirectional Associative Memories, *IEEE Trans. Systems, Man and Cybernetics*, Vol. 18, (1), pp. 49-60, 1988.

Levenberg, K., "A Method for the Solution of Certain Non-linear Problems in Least Squares", *Quarterly Journal of Applied Mathematics*, 2, 164-168, 1944.

Ljung, L., *System Identification: Theory for the User*, Prentice Hall, Upper Saddle River, NJ, 1987.

Marquardt, D. W., "An Algorithm for Least Squares Estimation of Non-linear Parameters", *J. SIAM*, 11, 431-441, 1963.

Masters, T., *Practical Neural Network Recipes in C++*, Academic Press, San Diego, CA, 1993a.

Masters, T., *Advanced Algorithms for Neural Networks*, John Wiley & Sons, New York, 1993b.

MATLAB, The MathWorks Inc., Natick, MA, 1994.

Miller, W. T., R. S. Sutton and P. J. Werbos (eds.), *Neural Networks for Control*, MIT Press, Cambridge, MA., 1990.

Mills, P.M., A. Y. Zomaya, and M. O. Tade, *Neuro-Adaptive Process Control*, John Wiley & Sons, New York, 1996.

Minsky, M., S. Pappert, *Perceptrons*, MIT Press, Cambridge, MA, 1969.

Moller, M. F., "A Scaled Conjugate Gradient Algorithm for Fast Supervised Learning", *Neural Networks*, Vol. 6, 525-533, 1993.

Moscinski, J., and Z. Ogonowski, Eds., *Advanced Control with MATLAB & SIMULINK*, Ellis Horwood division of Prentice Hall, Englewood Cliffs, NJ, 1995.

Narendra and Parthasarathy, "Identification and Control of Dynamical Systems Using Neural Networks", *IEEE Transactions on Neural Networks*, Vol. 1, No. 1, March 1990.

Nomura, H., I. Hayashi, and N. Wakami, A Self Tuning Method of Fuzzy Reasoning by Genetic Algorithms, in *Fuzzy Control Systems*, A Kandel and G. Langhols, eds., CRC Press, Boca Raton, Fl, 1994, pp. 338-354.

Omatu, S., M. Khalid, and R. Yusof, *Neuro-Control and its Applications*, Springer Verlag, London, 1996.

Park, J., and I. Sandberg, Universal Approximation Using Radial-Basis-Function Networks, *Neural Computation*, Vol. 3, 246-257.

Pham, D. T., and X. Liu, *Neural Networks for Identification Prediction and Control*, Springer Verlag, London, 1995.

Rosenblatt, F., "The Perceptron: a Probabilistic Model for Information Storage and Organization in the Human Brain", *Psychological Review*, Vol. 65, 1958, 386-408.

Shamma, S., "Spatial and Temporal Processing in Central Auditory Networks", *Methods in Neural Modeling*, (C. Koch and I. Segev, eds.), MIT Press, Cambridge, MA, 1989, pp. 247-289.

Specht, D., A General Regression Neural Network, *IEEE Trans. on Neural Networks*, Vol. 2, No.5, Nov. 1991, pp. 568-576.

Uhrig, R. E., J. W. Hines, C. Black, D. Wrest, and X. Xu, Instrument Surveillance and Calibration Verification System, Sandia National Laboratory contract AQ-6982 Final Report by The University of Tennessee, March, 1996.

Wang, L.-X., *Adaptive Fuzzy Systems and Control*, Prentice Hall, Englewood Cliffs, NJ, 1994.

Wasserman, *Advanced Methods in Neural Computing*, Van Nostrand Reinhold, New York, 1993.

White, D. A., and D. A. Sofge (eds.), *Handbook of Intelligent Control*, Van Nostrand Reinhold, New York, 1992.

Williams R. J., and D. Zipser, "A Learning Algorithm for Continually Running Fully Recurrent Neural Networks", *Neural Computation*, Vol. 1, 1989, 270-280.

Werbos, P. J., *Beyond Regression: New Tools for Prediction and Analysis in the Behavioral Sciences*, Ph.D. Thesis, Harvard University, 1974.

Werbos, P. J., "Backpropagation Through Time: What It is and How to Do It", *Proceedings of the IEEE*, Vol. 78, No. 10, October 1990.

White, D., and D. Sofge, Eds., *Handbook of Intelligent Control: Neural, Fuzzy and Adaptive Approaches*, Van Nostrand Reinhold, New York, 1992.

Zbikowski, R., and K. J. Hunt, *Neural Adaptive Control Technology*, World Scientific Publishing Company, Singapore, 1996.

CUSTOMER NOTE: IF THIS BOOK IS ACCOMPANIED BY SOFTWARE, PLEASE READ THE FOLLOWING BEFORE OPENING THE PACKAGE.

This software contains files to help you utilize the models described in the accompanying book. By opening the package, you are agreeing to be bound by the following agreement:

This software product is protected by copyright and all rights are reserved by the author and John Wiley & Sons, Inc. You are licensed to use this software on a single computer. Copying the software to another medium or format for use on a single computer does not violate the U. S. Copyright Law. Copying the software for any other purpose is a violation of the U. S. Copyright Law.

This software product is sold as is without warranty of any kind, either express or implied, including but not limited to the implied warranty of merchantability and fitness for a particular purpose. Neither Wiley nor its dealers or distributors assumes any liability of any alleged or actual damages arising from the use of or the inability to use this software. (Some states do not allow the exclusion of implied warranties, so the exclusion may not apply to you.)

WILEY